ROCKBURSTS
Global Experiences

Papers presented at the Fifth Plenary Scientific Session of Working Group on Rockbursts of International Bureau of Strata Mechanics, February, 1988

Organised by the
Institution of Engineers (India)

Editors
Prof. Ajoy K. Ghose
Prof. H.S. Seshagiri Rao

A.A. BALKEMA/ROTTERDAM
1990

ISBN 90 6191 140 0

FOREWORD

Since the beginning of mining, rockbursts have constituted one of the most serious hazards to man-made activities underground. Gradually, however, through practical experience and research work, many unexpected phenomena have been physically explained and methods of preventing damage to mine structures have been worked out.

At present, methods of stress measurement and analysis are available; the dynamics of rock failure process have been explained to a certain extent; seismic methods allow for the location of high stress concentration zones; and seismologic networks help to locate sources and calculate the energy of dynamic events in mines.

However, due to local variations of rock properties as well as the varied mining systems being used in various countries—the experience in preventing rockbursts is always unique and may be utilized by others only after careful consideration of the geologic and geomechanical environment. That is why an international exchange of information is so important, especially in the field of the prediction of rockburst hazard and preventing rockburst occurrences in mines.

The International Bureau of Strata Mechanics (IBSM), through its Working Group "Rockbursts", proposed an international seminar on rockbursts in India, cognizant of the long history of rockburst-related research and practical experience of Indian mining engineers and scientists, especially those involved in the mining activity at Kolar Goldfields. The Seminar was held in Hyderabad on 2nd and 3rd February 1988 and was followed by a visit to Champion Reef mine (KGF) at its deepest levels.

I am confident that the Seminar papers presented in this volume make a significant contribution to the knowledge of rockburst phenomena in mines and to the further development of rockburst prevention methods.

On behalf of all Bureau members and Seminar participants I congratulate the Indian organizers for their excellent arrangements during the Seminar and mine iv

visit. In particular, I acknowledge the perfect organization of the Institution of Engineers (India) and Bharat Gold Mines Ltd., for their action. Professor H.S. Sheshagiri Rao took responsibility for on-site organization of the Seminar and Professor A.K. Ghose for publication of the proceedings. Both of them have done their work at the top quality level—to the benefit of both the Seminar participants and to the readers of this volume.

PROF. DR. A. KIDYBINSKI
Chairman
International Bureau
of Strata Mechanics
(IBSM)

PREFACE

Worldwide, the mining industry has had to contend with the problems of rockbursts at depth since the beginning of this century. The problem dimension has exacerbated with increasing depth of exploitation and major research efforts have been underway to understand the mechanics of rockburst and the means and methods of combating the hazard. The Gold Mines at Kolar Goldfield, where working depths have reached over 3.2 km, have rich experiences of conditions that produce rockbursts. Significant research results at Kolar have also led to the control, by careful design of mine operations, of the frequency and magnitude of rockbursts.

This text includes a selection of papers presented at the 5th Plenary Scientific Session of the Working Group on Rockbursts of the International Bureau of Strata Mechanics held at Hyderabad in February, 1988 and marks a major contribution in synthesizing the existing state of knowledge on the occurrence, prediction and control of rockbursts, both in hard rock mining industry and coal mining. The meeting of the Working Group provided a unique opportunity to practical mining engineers to update their knowledge of rockbursts.

The Institution of Engineers (India), which is the Indian National Committee of the World Mining Congress, helped to provide the focus of efforts for the Fifth Plenary Session which attracted a distinguished group of experts, both from India and abroad, to present the results of their research and experiences. The meeting was possible largely through the initiative of Prof. A. Kidybinski, Chairman of the International Bureau of Strata Mechanics, to whom grateful acknowledgement is made.

We are confident that the publication of this volume will contribute to the search for improved methods of combating rockburst hazards. We warmly acknowledge the cooperation of the organisations which cosponsored this meeting.

PROF. AJOY K. GHOSE
PROF. H.S. SESHAGIRI RAO

July, 1990

CONTENTS

ROCKBURST PHENOMENA

1

THEORY OF ROCKBURSTS AND THEIR CLASSIFICATION

I.M. Petukhov

VNIMI, Leningrad, USSR

The problem of rockbursts in world mining science and practise is more than two hundred years old. Nowadays, however, this problem is real; the urgency of solving it grows steadily due to increasing depth of mining deposits, more extensive excavation volumes, improvement of mining engineering techniques etc. All these factors offer ample scope for progress in controlled safe utilisation of enormous reserve of potential energy stored in rock mass to facilitate mining and to reduce mining costs.

Soviet scientists were confronted with the problem of rockburst occurrences nearly four decades ago; they found themselves unprepared to meet these critical conditions, which prompted them to undertake comprehensive research into the problems of prediction and prevention of rockbursts. Since the fifties to the present research work has been carried out under the direct guidance of the All-Union Research Institute for Geomechanics and Mine Surveying (VNIMI). During this period, some of the problems concerning the safe mining of coal deposits have been practically solved. *Vis-a-vis* ore deposits, where investigations were begun only in the seventies, research work is under way on a range of advanced measures for rockburst prediction and prevention.

At the same time, it should be noted that the significant progress made in both coal and ore mining does not allow decelerating further research and experimental work aimed at ensuring safe working conditions for miners and increasing the economic efficiency of mining operations, profitability and production capacity, compared with mining deposits not prone to rockbursts.

The extensive experience gained in practise in recent years has confirmed that the application of a number of well-known measures may essentially somehow "reduce" the depth of deposit mining. This seems to be quite advantageous and, in certain cases, one may utilise the high stress of the rock mass for controlled deformation of rocks.

In the USSR, the problem of rockbursts is being tackled quite successfully, because from the very beginning much attention was paid to aspects of the theory of rockbursts, the problem of their prediction and prevention.

In the early fifties, a working hypothesis on the nature of rockbursts was submitted relating to the mechanism of occurrence of rockbursts and their energy balance [1]. According to this hypothesis, the entire system "coal seam-surrounding rocks" takes part in the causation of rockbursts. Energy balance of rockbursts consists of energy stored in the failed coal and of strain energy of surrounding rocks.

Just as the excessive inflow of energy from surrounding rocks induces the dynamic phenomena of rock failure, coal failure in the seat of a rockburst occurs due to exceeding the rate of pressure increase over the rate of stress relief at the cost of stress relaxation. This hypothesis appeared to be advantageous in the development of methods for rockburst prediction and for its prevention as well. It was also taken as a basis for the developed theory of rockbursts with due account of their diversity of manifestation in mining various mineral deposits [1–6].

This brief paper cannot describe in detail all the aspects of the theory of rockbursts. Let us consider only those that are of great importance for further development of this theory and for improvement of a package of measures for safe and efficient mining operations in deposits prone to rockbursts.

Development of the method of geodynamic zoning of deposits [4], on the basis of knowledge of science of earth and geomechanics, allowed mainly a different look at the rock mass within the deposit being mined. In view of the behaviour of the earth's crust, its block structure has been defined as well as interaction of blocks and finally the stress state of rock mass evaluated prior to mining operations. This explanation makes available a better understanding of the subject–geomechanics of rock mass. By "geomechanics of rock mass" we shall basically mean the study of mechanical and phase-physical properties and stress-strain state of rock mass with due consideration of the geodynamics of the earth's crust for providing safe and efficient mining, even with the existence of conditions for potential manifestation of rockbursts.

As a result of geodynamic zoning, the definition of block structure of rock mass, the nature and direction of block interaction, changes of stress state, detection of areas of rock tension as well as compression areas, and also the strain effects near marginal faults and rock disturbance in the region of deposit location—all these permit one to establish the boundary conditions to calculate the stresses in block rock mass using the method of finite elements or the method of boundary integral equations.

The normal practise in evaluating the stress state of rock mass according to Linnik, when the maximum principal stress acts vertically and is equal to γH, and horizontal stresses are equal to each other and are determined by lateral pressure, approximating to 1 with depth, may be used only when the given part of rock mass is in dilatation areas of the earth's crust.

In compression areas of the earth's crust, the maximum principal compressive stress σ_1 is pre-set by real force in the horizontal plane, and the principal compressive stress σ_3 coincides with force of gravity of rocks (γH). Another stress component in the horizontal plane σ_2 is determined by lateral pressure, induced by real force and gravity of rocks.

The average value of stress in compression areas of the earth's crust may be obtained from the ultimate stress state of rock mass. At comparatively low depths, an extremely discrete distribution of stresses takes place. It is caused by a number of factors, the principal one being: inhomogeneity of mechanical properties of rock mass (strength and deformation properties of lithologically similar rocks, variety of forms, dimensions of blocks, jointing, deposits, seams and their spatial relation), existence of tectonic stresses and stress relief caused by rock movement on irregular surfaces of contacts. It follows from this that if the averaged stress of rock mass is used in solving the problem of regional prediction of rockbursts, sub-dividing the deposit into mine fields and sequence of mining the coal seams or other deposits, then in solving the problems of rockburst prediction and prevention during drivage of mine workings and mining operations, one should take into account the assessment of stress state of rock mass obtained by instruments. In this case, it is not necessary to define the absolute values of stress state; it is quite sufficient to obtain only relative characteristics of stress state concerning the manifestation of some dynamic phenomena, excavation support etc. Geophysical methods of measuring the individual parameters for determination of rock mass behaviour are most advantageous. For deposits prone to rockbursts for example, geophysical methods and instruments have been developed, intended for regional and local prediction of rockbursts. They were applied in mines of production combines "Sevuralboxitruda," "Sibruda," "Kizelugol" and "Gruzugol," where deposits are highly prone to rockbursts. Here regional prediction is done by means of microseismological stations, and local prediction—by use of small, portable devices, which record high-frequency acoustic or electromagnetic emissions in the rock mass in the vicinity of mine workings. In the near future, the task will be to apply automated devices for continuous control of stress state of deposits prone to rockbursts and to use this system for regional and local prediction simultaneously.

In this connection the regularity of bump-like deformation of rock mass takes a specific place. This regularity means that changes of stress-strain state and energy, induced by mine workings in rock mass with discrete distribution of stress manifest in bumps, may be attributed to individual events of rock deformation (movements) caused by inflow (release) of energy from surrounding rocks, i.e., in conditions of "soft" loading.

Specifically intensive processes of bump-like deformation occurs in deposits located in compression and shear zones of the earth's crust, when changes of stress-strain state, induced by mine workings, arise from energy release from high strain acting in a horizontal plane. Block rock mass, adapting itself to a new stress-strain state in the form of bumps, tries to find stability in it. Investigations carried out in the Tkibuli-Shaorsk coal field, the Severouralsk bauxite deposit and the Tashtagol ore deposit showed that seismic energy of bumps reached 10^8–10^9 J, causing roof caving in poorly supported workings or occurrence of tectonic bursts there. The same effects can be observed in mining coal deposits prone to rockbursts in the presence of hard enclosing rocks. In this regard the explanation of the nature of rockbursts is somewhat enlarged.

Rockburst is the result of failure of a part of rock mass under ultimate strain conditions, induced by mine workings, when the rate of stress change in this area exceeds the rate of stress relaxation; energy of rock mass consists of energy stored in the failed rocks and of energy released from surrounding rocks. Here the conditions of "soft" loading are realised, i.e., the rate of energy inflow from the external medium exceeds the rate of its consumption during failure. This refers equally to rockbursts occurring in pillars and abutment areas of a rock mass, as well as rockbursts of the bump type occurring due to rock movements in a block rock mass on available or newly developed weak planes.

Let us cite some quantitative characteristics of hazardous areas of rock mass and of the rockburst itself.

First of all, the index of hazardous state of rocks (brittleness), reflecting the behaviour of the zone of rock mass under complicated strain conditions [1], is:

$$N = \frac{V_H}{V_p},$$

where V_H is the rate of change of stress state in the given area of a coal seam; V_p is the rate of stress relaxation in the same area.

Brittle fracture and, consequently, occurrence of rockburst is possible only when $N > 1$.

The index of rockburst dynamics may be:

$$K_d = \frac{W_{ext}}{W_f},$$

where W_{ext} is the energy entering from the external medium, when the rock mass has overcome compressive resistance, tension resistance, shear resistance or their combinations; W_f is the energy required for rock failure. The value of K_d can change from 0 to 10 and more.

The index of intensity of a rockburst is quantity of total energy participating in its occurrence. The intensity of rockburst occurrence may also be defined by a quantity of energy entering from the external medium. This index more adequately reflects the destructive force of a rockburst, i.e., its catastrophic failure potential.

In view of theoretical considerations of the nature of rockbursts as well as problems in their study, prediction and prevention, a classification of rockbursts has been proposed. According to this classification, the dynamic phenomena of rockbursts are sub-divided into five groups: rock outbursts, micro-bursts, bumps, rockbursts and tectonic rockbursts.

Rock outbursts—brittle fracture of rocks (ore, coal) on the exposed marginal areas of rock mass in the form of exfoliation and tearing of lens-like plates of various size with strong sound.

Micro-bursts—brittle fracture of edges of pillar or coal seam in the form of expulsion of rocks (ore, coal) into mine workings without heavy destructive after-effects and with no interruption in mining operations. Micro-burst is accompanied by sharp sound, weak shaking of the rock mass and formation of a dust cloud.

Bumps—brittle fracture inside the rock mass without rock failure in the vicinity of mine workings. Bump is accompanied by rock shaking and by sharp sound. Expulsion of dust is possible as well as breaks in concrete supports and rock caving in poorly supported areas.

Rockbursts—brittle fracture of a pillar, its part or a zone of rock mass in the form of expulsion of rocks (ore, coal) with harmful after-effects, interrupting mining operations. Rockburst is accompanied by strong sound, heavy rock shaking, severe dust dispersal and air blast.

Tectonic rockbursts—brittle fracture inside the rock mass in the form of bumps, inducing rock failure in pillars and seam edges adjacent to mine workings. A tectonic rockburst is accompanied by heavy rock shaking, strong sound, dust dispersal and air blast

A special Commission is engaged in investigating only two types—rockbursts and tectonic rockbursts. Since other categories of rockbursts characterise mainly the hazardous state of rock mass, within the deposit under exploitation, it is necessary to systematise rockbursts and to analyse them not less than once a month with a view to taking measures for their prevention.

Besides the above, there is also the classification of destructive and tectonic rockbursts according to loading conditions, defining their energy content [1–3].

The similarity in nature of rockbursts of various types and their energy properties would be more obvious if, on the one hand, one would regard the rock mass, within which mine workings are driven, as a single system having a certain strain energy, changes in which are induced by mine workings and, on the other hand, as a block system, with the strain state of each block depending on its spatial location in the rock mass and mode of interaction with adjacent blocks.

From this point of view, bumps and local earthquakes are of similar nature in hard rock extraction at great depth and in mining liquid and gaseous deposits and in the construction of hydro-engineering structures and their operational service.

Technogenic earthquakes, occurring in the earth's crust as a result of man-made activity, are induced by bumps; sometimes they can convert into tectonic rockbursts, which cause the destruction of mine workings and even in some cases damage of terrestrial structures and objects. Therefore, it is necessary to study more closely the manifestation of bumps in rock mass and to develop efficient measures for their prediction, for controlled bump-type deformation of rock mass, in order to ensure safe conditions for deposit mining and for operation of structures of various designations.

Solving the problem of controlled stress-strain state of a block rock mass must be based on established regularity of rock failure. This implies that with increase of rate of movement of rock exposure, the intensity of brittle fracture of rocks in the area near to face will rise, induced by concentration and localisation of rock pressure in that area. This regularity was also determined in the fifties by scientists of VNIMI for application to coal-seam failure caused by roadheading mining machines, mechanical cutting machines and rope saws, and in the seventies—in application to failure of sandstones with use of roadheading machines of types "Yasinovatets," KPT and "Sojuz-19".

Instrumented observations in coal and ore mines and tests of coal and other rocks with lever press and rigid press showed that main processes of stress relaxation in rocks occur immediately in the first minutes after mechanical cutting. It means that cutting speed of mining machines is comparable with rate of redistribution of rock pressure. Application of universal mining and entry-driving machines, rope saws on the basis of regularity of rock failure and advisable use of rock pressure showed that some operating characteristics of mining machines may be improved, i.e., power-to-weight ratio may be considerably reduced. Mining machines with appropriately chosen parameters of control member and operational regime could only control the process of rock deformation, providing for the safe and efficient use of rock pressure.

Further development of the theory and techniques of controlled deformation of coal, ore and other rocks, with automatic control of rate and intrusion parameters in drivage of mine workings and in actual mining, may be quite advantageous and effective and yield economic benefits.

In recent years, scientists of VNIMI have been engaged in the development of a unified theory of rockbursts and outbursts of coal and gas, and also of a package of safety measures for efficient mining of deposits prone to rockbursts and outbursts [1, 2].

First of all, these aspects are related to the necessity for proper division of the deposit into mine fields, sequential order of their mining and providing the elimination of excessive concentrations of rock and gas pressure in the vicinity of mine workings.

Unified perspective schemes of regional control of rock and gas pressure for coal seams prone to rockbursts and outbursts of coal and gas are under study now. They will be used in all coal mines in the USSR. The application of unified perspective schemes in practice will allow the use of regional measures for predicting rockbursts and outbursts in mining coal seams in all coalfields and eliminate the use of expensive and inefficient local measures.

The development of a unified theory of rockbursts and outbursts and improvement in methods for their prediction and prevention seem to be the principal trends in mining science and in practise in the foreseeable future.

The safe and efficient mining of hard, liquid and gaseous deposits at great depths will be realised on the basis of comprehensive study of mechanics and gas and hydrodynamics of block rock mass with obligatory use of geodynamic zoning of deposits.

REFERENCES

1. Petukhov, I.M. 1972. *Rockbursts in Coal Mines*. Nedra, Moscow.
2. Petukhov, I.M. and A.M. Linkov. 1963. *Mechanics of Rockbursts and Outbursts*. Nedra, Moscow.
3. Petukhov I.M., P.V. Egorov and B.Sh. Vinokur. 1984. *Prevention of Rockbursts in Ore Mines*. Nedra, Moscow.
4. Batugina, I.M. and I.M. Petukhov. 1988. *Geodynamic Zoning of Deposits in Designing and Exploitation of Ore Mines*. Nedra, Moscow.
5. Petukhov, I.M. and V.A. Smirnov. 1977. *Geophysical Investigations of Rockbursts*. Nedra, Moscow.
6. Petukhov, I.M. 1987. Classification of rockbursts, *Safety Operations in the Mining Industry*, No. 12. Nedra, Moscow.

2

INVESTIGATION OF THE MECHANISM OF ROCKBURSTS BY SEISMOLOGICAL MEASUREMENTS

P. Knoll and W. Kuhnt

Central Institute for Physics of the Earth, Potsdam, GDR

INTRODUCTION

In recent years, mining and geomechanical investigations have shown that there are two main types of rockburst mechanism. One has to use different methods of prevention and control in mines for each type. Therefore, a detailed knowledge of the mechanism and reliable classification of the different types of mechanism are of great importance in burst-prone mines. Through seismological measurements and application of simple source models, focal parameters such as seismic moment, source radius, stress drop and dislocation in the source can be estimated; scaling laws of these source parameters allows the classification of bursts into main types. More complicated source models for a known source mechanism allow calculation of parameters giving an insight into the degree of destruction within the source region underground. The advantage of seismological investigation is the possibility of estimating source parameters with methods acting far off the fracture zones, without disturbing the technological process and giving electrical signals sufficiently good for computer processing. This is demonstrated by some examples in this paper.

THE MECHANICS OF ROCKBURSTS

Conception of Rockburst Control

The occurrence of rockbursts in all branches of underground mining as well as in tunnelling and sometimes even in near-surface extractions is very old and even today not a fully solved problem of mining rock mechanics. Therefore, the need to find the causes of rockbursts and to derive from them measures to control the rockburst menace is nearly as old as mining itself. Meanwhile, diverse methods have been tried with varying degrees of success to predict, combat or prevent rockbursts.

Rockburst experiences in mining practise show the occurrence of very different kinds of rockbursts related to very different effects on underground openings and on the surface near mines. According to the different kinds and effects of rockbursts, there are different designations and definitions. Likewise, mining practise shows, that it is impossible to combat the occurrence and danger of all types of rockbursts by using the same methods. One of the most important steps in rockburst control is therefore the investigation of fracture process in the sources of bursts, i.e., analysis of the physical reasons promoting such conditions and natural and man-made prerequisites.

The goal of this investigation must be the classification of fracture process within the source into main types of focal mechanism. Classification in relation to the mechanism is much more effective for determining the right countermeasures against rockbursts and for differentiating the effects or damages respectively, within mining openings.

The next step is to find relations between the main types of rockbursts and mining activities or natural conditions respectively by theoretical analysis and experimental investigations as well as measurements and observations. On that base, more certain and more exact conditions are possible for prediction, control and limitation of effects. It is very important to check the efficiency of the different measures chosen and applied in the mine to control rockburst danger for different types and to draw conclusions for better classification, better modelling of the focal process and better countermeasures. Figure 1 shows the essential steps in the conception of rockburst control based on rock mechanical and seismological investigations and measurements. Some results and examples of investigations of rockbursts following this conception will be presented below.

Types of Rockburst

Analysis of extensive investigations of the occurrence and effects of rockbursts and seismic events in mining areas in several countries, including the GDR, as well as geomechanical and seismological analysis of the rockburst mechanism, have shown that in sudden fracture processes of rock in mining areas there are two fundamental mechanisms that differ essentially from each other (Knoll, 1981; Knoll, Thoma, Hurtig, 1980).

Experience in combating rockburst danger and simple rockburst statistics

CONCEPTION FOR ROCKBURST CONTROL

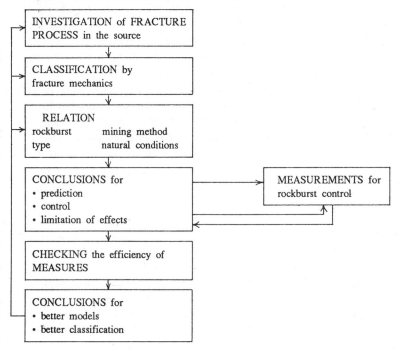

Fig. 1. Conception of rockburst control based on rock mechanical
and seismological investigations.

support classification into two main types. For example, in Indian gold mines of Kolar Goldfields local rockbursts and so-called "major area rockbursts" have been observed (Krishnamurthy, Nagarajan, 1983), in the Caucasian coal deposit in Tkibuli, USSR, (Smirnov *et al.*, 1973) local rockbursts as well as regional mining-tectonical ones, and in the deep South African gold mines small induced seismic events occurring within a short distance from the working face and large induced seismic events with a magnitude greater than 1.5 in the Richter scale (Salamon, Wagner, 1979). It is particularly clear in the Western Polish copper mines, for example, that the local rockbursts near the working face and the mining-induced event of March 24, 1977 (Gibowicz *et al.*, 1979) are quite different vis-a-vis the physical mechanism and, of course, the effects in the mine. Some theoretical reflections on this topic are also given by Shemjakin (1987).

Following Knoll (1981), the two main types of rockburst can be characterised as follows:

Type 1 (called mining or static rockburst) is closely connected with face mining. Therefore, its foci are located in the direct vicinity of mining openings, mostly in their perimeter areas (pillar edges, face advances). From the point of view of rock mechanics these rockbursts are bound to zones which, as a direct consequence of stress redistribution in connection with the advancing face, are highly loaded and thus reach almost the critical state. Different triggering mechanisms (in most cases face blasting) can then trigger the brittle fracture connected with energy release. The foci of these fractures have been seismically recorded as located near active working faces. Geometrical extensions of focal zones correspond to zones of stress redistribution affected by respective local mining activities. Further parameters characterising the focal process can be derived from mining parameters, taking into account the geomechanical, geological and tectonic conditions of the mining field. The stiffness distribution in the rock of the focal zone is of great importance in assessing released energy.

Investigations have shown the expediency of distinguishing two modifications of Type-1 events.

Figures 2a and 2b schematically depict possible focal processes according to Type 1. Type 1a shows the actual overloads of perimeter zones. Destructions occur only in the very local region represented by highly stressed perimeter zones. Type 1b shows the development of focal areas within rock formations in the immediate roof or floor rock strata where there are increased shear stresses due to differing stiffness of the rock massif (e.g., between the solid virgin and excavated deformable parts). In that case, energy emission will occur on the shear zones within the hanging or footwall strata respectively, and the damages in the openings will be a follow-up of that emission of seismic energy into

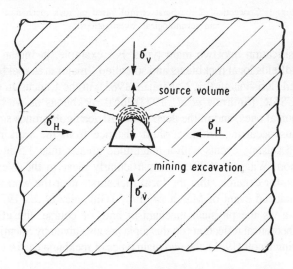

Fig. 2. Scheme of fracture mechanism of Type 1 rockbursts:
(a) parameter bursts.

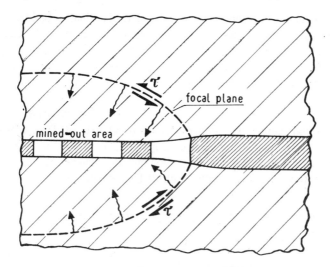

Fig. 2b. Near opening burst.

openings near the focal zones (within the working space influenced by the focal area) if the conditions of the occurrence of brittle fractures induced by dynamic loading are given at the perimeters of the openings. Also Type 1b events are related to the rock mass near the working faces because of the maximum stiffness differences in these zones and the missing mechanical equilibrium.

Types 1a and 1b, of course, can occur not only in single roadway (Fig. 2a) or room and pillar mining system (Fig. 2b) respectively, but also in all other openings, opening systems and mining methods.

Type 2 (called tectonic or dynamic rockburst) is only indirectly connected with mining activities. Its formation results from regional stress redistribution within the rock mass in the far field environment of the stoping areas. Stress redistribution is due to mining activities at large working areas or whole mines. This type is triggered by high tectonic stresses already in the virgin rock massif where they occur predominantly at great depths. The bursts originate especially when geological faults are orientated appropriately to the tectonic stress field in the form of zones of weakness capable of brittle fracturing.

A further prerequisite is that rock deformations due to mining reach a value between the fault zones large enough to cause important stress redistributions in a large rock volume but not so large, however, as to result in a disintegration of the rock mass or passing over to a post-failure state within weakness zones. This condition becomes increasingly important with the utilisation of low-loss mining methods. Depending on the kind of faults, the critical state can be reached with only relatively low additional stresses. Of course, not all kinds of geological faults tend equally to act as focal regions for seismic events. This trend depends

16

on their geomechanical properties and the deformational properties of the environment of the potential focus (Knoll, Schwandt, Thoma, 1979). The character of Type-2 rockbursts leads to the further conclusion that they occur less frequently than Type-1 rockbursts, as the necessary regional stress redistributions are the result of long-time mining processes over extensive areas.

On the other hand, compared with Type 1, considerably larger rock masses reach the critical state and higher rates of energy are released during fracturing. There is a special danger for mining because rockbursts of Type 2 need not necessarily be triggered by mining (blasting) and, due to higher magnitudes, there arise higher dynamic loads on the mining openings and supporting elements. The effect of these loads is hardly controllable (Knoll, 1979a).

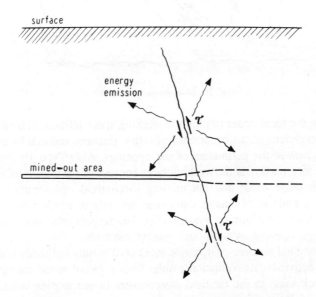

Fig. 3. Scheme of fracture mechanism of Type 2 rockbursts.

Figure 3 shows the principal characteristic features, the long distance between the focus of energy release on a tectonic fracture zone within the rock mass and the location of the effects, of fracturing in mine openings. This sketch also demonstrates that the burst not only affects the active mining faces but also vast areas of the exploited mine. The destructions are partly locally isolated and seem to be completely independent of each other, but cover a large area (like major area rockbursts in Kolar Goldfields). This depends on whether the prerequisites for fracturing engendered by dynamic loading are given at the respective places. These features are the cause of the high danger of rockbursts of Type 2 and the difficulties of carrying out effective countermeasures from the working face, as these do not reach the actual focal zone.

Rockbursts of Type 2, first explicitly mentioned by Knoll (1981) and Knoll *et al.* (1980) have subsequently been noticed and described in several mining regions (Gendzwill *et al.*, 1982; Kosyrev, 1985; Wong, 1985).

Brief characterisation of the geomechanical features of Type 1 and Type 2 events shows:

— Classical methods to combating rockbursts, as for example, stress-relief blasts, water infusion, stress-relief drilling and others, as well as classical prediction methods such as core discing etc., are limited in their efficiency to Type 1a, events.

— Countermeasures can have a certain effect on Type 1b too by changing rock previously capable of brittle fracturing in the perimeter area into a non-brittle-fracturing pre-destructed state. Thus these countermeasures reduce the *effects* of rockbursts. However, they cannot prevent the fracturing process as such within the focus or the release of energy. Under certain conditions, the above-mentioned countermeasures even favour the Type 1b bursts, if the measures reduce the stiffness of the rock in the mined-out part and thus lead to increasing shear stresses in the focal region.

— Type 2 events, however, cannot be influenced by local measures in the perimeter area. They can only be influenced by appropriate mining strategies which properly take into account the tectonic structure of the rock and its stress conditions. The goal of these strategies has to be limitation of shear stresses on certain tectonic fracture zones.

As general laws for the occurrence of Types 1 and 2 cannot be given due to the great variety of natural and technical conditions in different mines and deposits, measuring methods (such as seismological investigations) for the classification of rockbursts within the given categories are relatively important for each rockburst-prone mine.

SEISMOLOGICAL INVESTIGATIONS

Methodology

One part of seismological investigations includes methods for localisation, energy estimation, calculation of magnitudes and statistics in time. These investigations are necessary and well known in all seismologically controlled mines.

The second part is more concerned with physical parameters of the fracture process within the source. For classification purposes, this part of the investigations is more important.

Seismic focal processes are described by the common geomechanical model conception of a brittle shear fracture along a more or less plane weakness

18

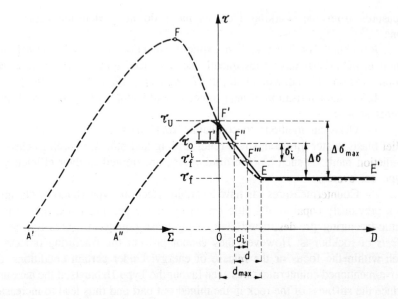

Fig. 4. Complete stress strain curve for monolithic rock (A″FEE′) or a weakness zone in the rock mass (A″F′EE′).

zone in the rock mass. Figure 4 schematically presents a complete weakness zone deformation curve (A″ F′ EE′) demonstrating the principal factors of influence. Let stress level τ_o exist before fracturing and level τ_f^i be the result of the fracturing. In extreme cases, τ_f^i can become τ_f if the stiffness of the focal vicinity is smaller than that of the focal volume itself. As τ_o can reach τ_u as a maximum, we get maximum values for the stress drop $\Delta\sigma$ to $\Delta\sigma_{max}$ $\geq \Delta\sigma \geq \Delta\sigma_i \geq 0$ and $d_{max} \geq d \geq d_i \geq 0$ in dependence on stiffness conditions in the focus and its environment. Mostly, in seismic models, the actual run of a complete deformation curve of an area of weakness (A″F′EE′) is replaced by a curved line TT′F′EE′ formed by several straight lines. The better the quality of the seismic model conception, the more exactly the approximation curve can be adapted to the actual run of the curve of a natural weakness area, the more exactly the conception of a shearing process corresponds to the actual geomechanical fracturing process within the source and the more exactly the fracturing process of the focal area can be described by a so-called focal function Ω:

$$\Omega\left(x_i, t\right) = \mu \int_s ds\left(x_i,\ t'\right) \Delta U_i\left(x_i', t - \frac{|x_i - x_i'|}{c}\right) \tag{1}$$

with μ – shear modulus;
   U – dislocation function;
   x_i – focal area;
   c – velocity of wave phase under investigation.

This yields the functions of the real movement of the ground in the time domain:

$$U(x_i, t) = (4 \, \eta \, \rho \, c^3 \, [x_i])^{-1} \, k. \, \Omega(x_i, t) \tag{2}$$

with ρ – rock density;
k – local wave emission characteristic;
and in the frequency domain:

$$U(x_i, \omega) = (4 \, \eta \, \rho \, c^3 \, [x_i])^{-1} \, k_i \, \omega \int\limits_S ds(x_i) \, e^{-i\omega \frac{|x_i - x_i'|}{C}} . \Delta U(x_i, \omega) \tag{3}$$

respectively.

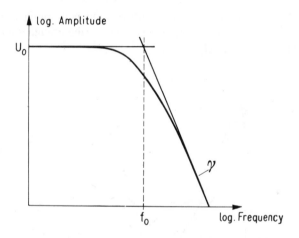

Fig. 5. Typical amplitude spectrum of seismic signals (schematic).

The typical form of this function is shown in Fig. 5. It contains three characteristic values for the respective focal processes:
u_o – so-called low frequency level;
f_o – corner frequency;
γ – high-frequency decay.
After spectral analysis of seismic records of focal processes, these values can be determined for any event without using special geomechanical focal conceptions.

Then, a model of the focal fracture process is used, which often consists of a round focal area with the fracturing starting in its centre, propagating with a finite propagation velocity v_B and stopping after having reached the focal radius r_o.

20

Such a model conception after Madariaga (1976) using the propagation velocity of the fracture $v_B = 0.6\ v_m$ (v_m-shear wave velocity) is, according to Kuhnt (1985), especially suitable for mining events. With the help of this model, different and partly complicated geomechanical focal processes can be transformed into a simple uniform model. From model-independent values u_o, f_o, γ the partly model-dependent values as seismic moment M_o, focal radius r_o, stress drop in the source $\Delta\ \sigma$ and dislocation on the focal area d can be calculated. This way, the different seismic processes become comparable from a quantitative point of view. Their classification is possible then by classification of scaling laws of focal parameters for the same source model and, indirect conclusions about the geomechanical process using the relations between the parameters and their geomechanical interpretation are also possible.

Furthermore, by comparison of the corner frequencies of the p-and s-waves ($f_{o,p}$ and $f_{o,m}$, respectively), one can roughly evaluate the orientation of the focal zone in the space respectively.

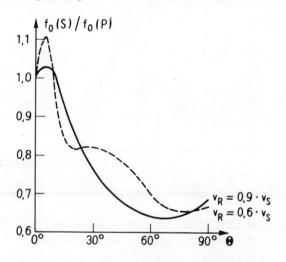

Fig. 6. Corner frequency ratio f_oS/f_oP as a function of the dip angle of the focal plane Θ for the Madariaga model (Kuhnt, 1985).

Figure 6 shows the correlation between $f_{o,m}/f_{o,p}$ and angle Θ formed by the dip direction of the focal area with the vertical (Kuhnt, 1985). Even though this dependence is relatively complicated, we can at least see that with $f_{o,m}/f_{o,p} \geq 1$, the focal area is steep and with $f_{o,m}/f_{o,p} < 0.7$ the focal area lies more or less flat.

Another way to estimate the spatial orientation of the focal zone is given by Kisslinger et al., 1982 and applied to rockbursts in potash mines by Behrens et al., 1987.

For many practical cases, such findings can give important hints about the focal process and the possible participation of tectonic elements in it.

Results Using the Simple Uniform Shearing Model

Focal parameter determinations in several mining areas given in the literature, when calculable for the uniform model given by Madariaga, were compared with each other and with the respective parameters of a weak tectonic series of earthquakes.

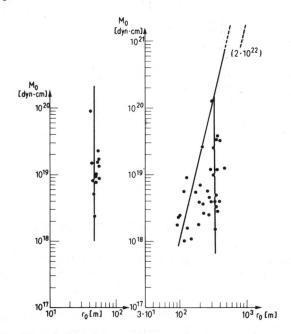

Fig. 7. Scaling laws M_o/r_o for rockbursts: (a) hard coal mining in Ruhr district/FRG (data: Hinzen, 1982); (b) copper mining in Western Poland (data: Gibowicz *et al.*, 1979, 1980).

Figures 7, 8 and 9 show single results for the correlation $M_o - r_o$. It is obvious that a group of events is characterised by nearly steady focal radii r_o and by seismic moments being variable over more than two orders (Figs. 7a, 8b). A second group is characterised by increasing M_o with increasing r_o (Fig. 8a). The kinds of events forming this group dominate either in special types of deposits or occur together with the first-mentioned type (Figs. 7b, 9). The practically steady geometric conditions (r_o const.) of the first group hint at the decisive function of technological mining parameters such as face width, room or roadway width, pillar dimensions etc. This expresses a direct relation

22

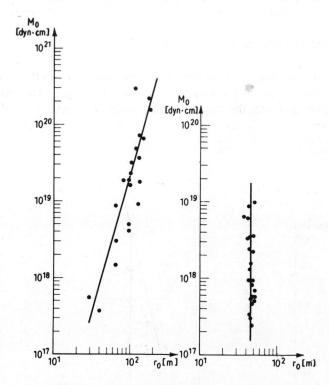

Fig. 8. Scaling laws M_o/r_o for rockbursts: (a) hard coal mining in Upper Silesia/Poland (data: Gibowicz *et al.*, 1977); (b) hard coal mining, Sunnyside Mine Utah/USA (data: Smith *et al.*, 1974).

between source radius and mining process, which is typical for rockbursts of Type 1. Obviously, in the second group the direct influence of mining activities decreases. The events are triggered by mining but then determined in their process by other (natural tectonic) factors. It is apparent that they be related to rockburst Type 2.

Figure 10 presents the compilation of results and a comparison with a series of tectonic earthquakes. Besides that, the lines for equal stress drop are given. The second group of mining-induced seismic events shows a correlation of $M_o - r_o$ similar to tectonic quakes. As tectonic quakes are determined only by the natural geologico-tectonic conditions in the earthquake focal area, one can conclude that this fact is expressed by the characteristic run in the $M_o - r_o$ scaling law.

So the mining-induced events of the tectonic type are likely to be correlated with tectonic features (fault zones, stress state) in the environment of the mine and are simply induced by mining activities.

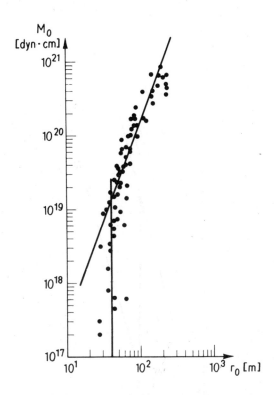

Fig. 9. Scaling law M_o/r_o for rockbursts in South African gold mining, Witwatersrand (data: Spottiswoode, McGarr, 1975).

The interpretation of the stress-drop parameter $\Delta\sigma$ (stress drop in focus during the seismic event) leads to similar conclusions. The Type-2 scaling laws as well as the scaling laws for natural events show a nearly constant stress drop (i.e., the straight lines $M_o - r_o$ run approximately parallel to the lines of equal stress drops in Fig. 10). This becomes quite understandable when we assume that such natural tectonic conditions as stress state, geomechanical fault parameters and spatial fault distribution (and thus geomechanical stiffness distribution in the rock) in a much wider area are nearly constant, compared with the extensions of underground openings. Figure 10 shows that Type-2 events in deep gold mines in South Africa ($M_o > 10^{19}$ dyn cm) with high values for rock strength have very high stress-drop values of about 100 bar. They have quite the same values as those of the San-Fernando earthquake series. In Polish copper mines, mining depths are only 1/3 those of gold mining in South Africa; the stress-drop values are about two orders lower. Even lower are the values for Polish hard coal mining with smaller depths and lower rock strength. These

24

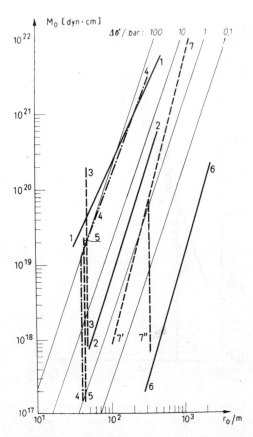

Fig. 10. Compilation of scaling laws (Figs. 7, 8, 9) together with scaling
laws for weak earthquake series: 1–San Fernando earthquake series, 2–Mijako
micro-earthquake series, 3–Ruhr district (Fig. 7a), 4–South African gold mining
(Fig. 9), 5–Utah (Fig. 8b), 6–Upper Silesia (Fig. 8a), 7–Western Poland (Fig. 7b).

characteristic stress-drop values for Type-2 events in certain mining districts
are nearly constant for quite different seismic moments. On the other hand,
at the working faces and other openings geologically and technologically
conditioned relations act which can result in different rock-stiffnesses and thus
in different stress-drop values for Type-1 events.

These correlations are to be discussed and examined using Fig. 4. If rockbursts of Type 1 occur, e.g., in a room-and-pillar stoping system with remaining
pillars, then it is geomechanically important for equal r_o (corresponding, e.g.,
to the pillar dimension or the dimension of a group of pillars) whether only
the highly loaded perimeter zones of the pillar are breaking and thus become
the focus of a rockburst, or whether the whole pillar or group of pillars respectively

are completely involved in the rockburst and collapse. In the case of perimeter burst the triaxially fixed stable core of the pillar remains. It has high stiffness which keeps the dislocation within the focus at a limit and the stress-drop value too. For Fig. 4 this would mean that the geomechanical state on curve A''F'EE' moves from F'' to F''' and dislocation d_1 and stress drop $\Delta \sigma_1$ arise. With complete failure of the pillar, state F' and F'' respectively would sooner reach state E, because the stiffness of the local environment is significantly lower due to the broken core of the pillar. Dislocation and stress drop would approximate the values d and $\Delta \sigma$ or d_{max} and $\Delta \sigma_{max}$ respectively. Both fracturing processes would have approximately the same focal dimension r_o (corresponding to the pillar dimensions), but would essentially differ in values of stress drop $\Delta \sigma$ (and seismic moment M_o).

Analogously, local differences in rock stiffness may occur with other mining methods too, due to different load behaviour of rock masses, different effectiveness of pillars, different filling behaviour etc. That can be reflected in different values for M_o and $\Delta \sigma$ for the rockburst.

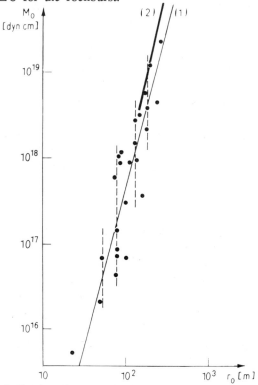

Fig. 11. Scaling laws for a potash mine with room and pillar stoping system: 1—mean curve for Type-1 events, consisting of several short typical Type-1 scaling laws; 2—Type-2 scaling law.

In Fig. 11 are shown the results of seismologic investigations of rock-bursts in a potash mine of GDR, adopting stoping by room and pillar technology. The pillars are more or less uniform and stable but as a result of the great depth (to 1,000 m) and the low strength of the rockburst-prone brittle potash-salt carnallite ($\sigma_c \approx 10$ MPa) the pillar edges are overloaded and Type-1 events (1a and 1b) by brittle fracturing of the pillar edges can occur (Behrens et al., 1987). Under certain conditions, tectonic rockburst (Type 2) in principle can also occur (Knoll, 1979b).

The scaling law for Type-2 events (2) in Fig. 11 has the same shape as shown for other deposits in Figs. 7b and 8a. But in Fig. 11, one can only see the lower part of the curve (2). Type-2 events have larger M_o values, as shown in Fig. 11. The scaling law for Type-1 events shows some short single curve parts (broken lines) with steady r_o values corresponding to the dimensions of one, two, four, nine etc. pillars. Because of the uniform deformation and strength characteristics of the pillars there are no dominant large-scale geometric parameters resulting in constant values of r_o for a broad scale of seismic moments. The short-curve parts with constant r_o gives an average line (1) looking like the curves for Type-2 events (2) with a smaller gradient (see Knoll et al., 1984). The scaling law over the full M_o scale looks like that for South Africa (Fig. 9) with Type-2 events for high M_o and Type-1 events for low M_o values, with the above-discussed special shape of the curve (1).

Other Seismological Models

The Madariaga model for fracture mechanism in the source is one of the simplest. It assumes a homogeneous fracture process over the entire source plan. But in situ observations in a mine show that the fracture within the source is much more complicated. Focal parameters such as, e.g. $\Delta \sigma$ estimated by the Madariaga model, are mean values over the whole focal plane, different from the true $\Delta \sigma$ value within the really broken parts of rock within the focal plane.

One possible way to better describe the fracture process of a rockburst is to introduce more complicated source models. But a prerequisite of that is improved knowledge of the type of rockburst.

Assuming that a rockburst is of Type 1 (Type 1b, e.g.), one can use models with an inhomogeneous stress drop. The barrier and the asperity models are one step in this direction (Das, Aki, 1977; Das, Kostrov, 1983; Das, Kostrov, 1985; Aki, 1979).

In terms of the barrier or the asperity model, the source process of a rockburst of Type 1 in a room-and-pillar mining system could be described as follows (Fig. 12):

Fig. 12a. Source plane with barriers or asperities (hatched area).

    a) **Barrier model**: The crack begins at the stope and propagates to a region near the caving zone. The large pillars act as barriers with higher strength and the crack stops without breaking the 'barriers'. The effective source area is comparable to the area of a caving zone. In Fig. 12, the caving zone is the broken part of the source plane.

    b) **Asperity model**: There is already a zone of weakness in the source region and the pillars (asperities) act as stress concentrators. With an increase in tectonic or mining-induced stress level, the asperities break (hatched area

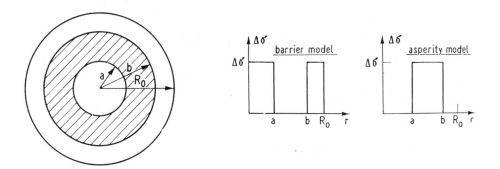

Fig. 12b. Left: Approximation of barrier model and asperity model by annular-shaped areas (hatched area—total area of barriers or asperities); Right: Stress-drop radial coordinate r in the source plane.

in Fig. 12). The stress drop is concentrated on the asperities but a dislocation appears also outside the asperity. At the end of the weakness zone the process is stopped by a material with higher strength.

Kuhnt et al., calculate on the base of these models the ratio $\Delta\sigma_{app}/\Delta\sigma$ for simplified asperity and barrier models of rockbursts (Type 1) in potash mines using room-and-pillar mining. The really broken part of the source plane is given in this analysis by the radius r_{cr} (Kuhnt et al., 1987). $\Delta\sigma_{app}$ is the 'apparent' stress drop resulting from the Madariaga model. It is not true, however, because calculated for the full focal plane with broken and unbroken parts. $\Delta\sigma$ is a truer value because only calculated for the broken parts of the focal plane (using the simplified asperity model).

In most cases of rockbursts of Type 1, the area of destruction can be estimated. In general, this area is not coherent. Therefore, the radius of the corresponding total area can be calculated by adding the single areas. The radius of the resulting area is supposed to correspond to r_{cr}.

The results show that the nearly constant $\Delta\sigma$ can be considered a parameter which characterises a certain source region. Supposing this $\Delta\sigma$ is the maximum stress drop $\Delta\sigma_{max}$ of this source region, the ratio r_{cr}/r_o of rockbursts with unknown r_{cr} can be calculated for a special simplified case as follows:

$$\left(\frac{r_{cr}}{r_o}\right)^2 = \varepsilon = 1 - \left(1 - \frac{\Delta\sigma_{app}}{\Delta\sigma_{max}}\right)^{2/3} \tag{4}$$

with $\Delta\sigma_{app} = \frac{7}{16}\frac{M_o}{r_o^3}$ and $r_o = \frac{kv_s}{f_o}$ $\qquad(5)$

Here the parameter k is taken from the Madariaga model (Kuhnt et al., 1984). It was found that the ratio r_{cr}/r_o is closely connected with the degree of damages of pillars or of faces in the mine. Figure 13 gives a quantitative relation between the ratio r_{cr}/r_o and the observed forms of destruction. It is shown that the value r_{cr}/r_o can be used for a rough classification of the degree of the damage. Characteristic values which depend on the considered source region are observed for the damage types "destruction (collapse) of pillars" (G1), "destruction of pillar edges" (G2) and "no damages" (G3). So one can get a first estimation of the effect of rockbursts in a mine using seismological information without in situ inspections, if one has estimated beforehand the characteristic value $\Delta\sigma_{max}$ of a certain mining region.

EXPERIMENTAL RESULTS IN A POTASH MINE

In a potash mine in the south-western part of the GDR within a tectonically complicated situation on the tip of the South German Block (Fig. 14), tectonic (Type 2) as well as mining (Type 1) rockbursts can occur.

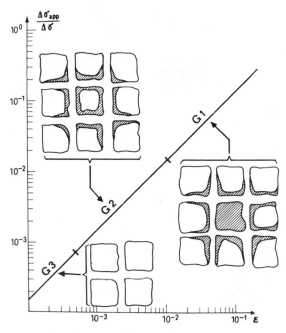

Fig. 13. Relationship between the ratio $\Delta\sigma_{app}$ $\Delta\delta$ and ε ($\varepsilon = (r_{cr}/r_o)^2$) and its qualitative relation to the damage in a mine.

The mining method is room-and-pillar mining in a flat deposit and two levels. The depth varies from about 400 m to about 1,000 m. One of the stoped potash salts, carnallite has a low strength (uniaxial compression strength $\sigma_c \approx$ 10 MPa) and is a very brittle rock. The edges of pillars consisting of carnallite at depths more than 700 m can break in Type 1a rockbursts and near the stoping faces Type 1b rockbursts can also occur. The mine is controlled by geomechanical and seismological measurements (Rösing et al., 1987).

The conception of seismological mining control consists of a wide array of seismometers whose signals are recorded automatically, if a triggering criterion is fulfilled. Figure 15 shows the distribution of seismometers in the mining region. Seismic signals are measured on a very wide band. The first frequency ranges from 2 to 100 cps and is covered by velocity seismometers. The second frequency ranges from 2 to 2000 cps; these are acceleration seismometers. Experience has shown, that the conception of a large frequency width is very successful in the solution of fracture problems. Velocity seismometers are used to measure events with a stronger release of seismic energy. Smaller events relevant to mining control (e.g., gas outbursts) are measured by acceleration seismometers. Additionally, one seismometer of the centralised seismic network

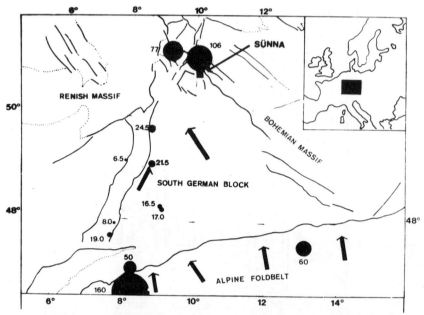

Fig. 14. Tectonic map of central Europe with horizontal stress excess (full circles). Mining area situated on the tip of the South German Block.

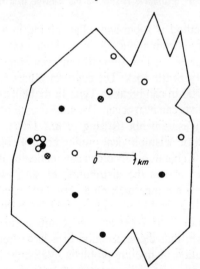

Fig. 15. Distribution of seismometer in the mining region. Open circles: position on the second underground level; full circles: position on the first underground level; crossed circle: position on the surface (Behrens *et al.*, 1987).

of the GDR Central Institute for Physics of the Earth is situated underground in the mine.

Rockbursts of all types are recorded and analysed by spectral analysis and the focal parameters determined. Very interesting is a comparison between the source radius r_0 and the situation in the mine after the event. A good correspondence between r_0 and the area with damages on the pillar edges for Type-1b events was often found.

Figures 16 and 17 present two examples. R_0 (P) and R_0 (S) are the source radii, calculated using p- or s-waves respectively. $\bar{R}_0$ is the average. A circle with the radius R around the location point of the source (near blasting point) includes the damaged zone in the mine very well.

Another example is given by Behrens *et al.*, 1987. Here the perimeter zone of a special large and high opening was damaged by a rockburst. Analysis showed that the path of the source region in a cross-section of the deposit in Fig. 18 started with an angle of about 45° in the hanging wall (maybe also in the footwall) strata and moved away from the virgin part of the deposit.

Table 1 shows the single shocks within about a 15-min long seismic event. Figure 18 shows the detailed localisation of the shocks and the depth position roughly. The calculated spatial orientation was:

Strike : N 20°E
dip : 45°W   dip-slip-mechanism.

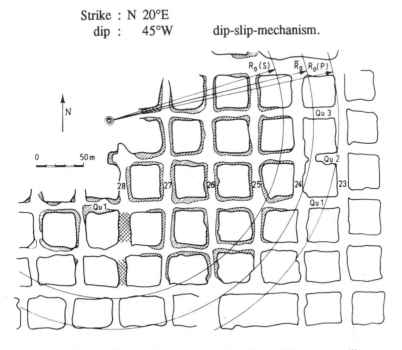

Fig. 16. Comparison between determinations of the source radii.

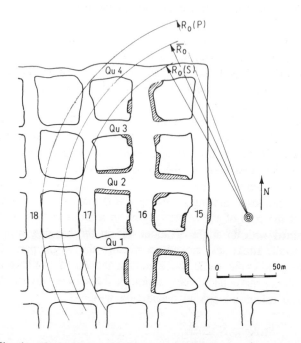

Fig. 17. Situation after seismic events for two examples. Dashed areas broke as a result of the events, crossed circles were the blasting points, and double circles are approximate source locations.

About a Type-2 event with a Richter magnitude of 5.4 and a destroyed underground area of 3.4 km² in this deposit was reported by Knoll (1979b, 1981). Hurtig *et al.* (1982) have shown interesting results of theoretical investigations of the fracture mechanism of that event, especially comparisons between synthetic and recorded seismograms. The results obtained in this potash mine are very good confirmations of the existence of rockburst Types 1 and 2.

CONCLUSIONS

Knowledge of the existence of two principal types of rockbursts—mining and tectonic rockbursts—and evidence for them in mining practise as well as their quantitative characterisation by focal parameters by means of seismological methods, is of great importance for practical control of rockbursts. Seismic monitoring of mines provides information on which type of rockburst prevails in which sections of deposits and for what kind of mining activities. The most appropriate measures taking into account the actual focal process can be considered for the control of rockbursts. According to the actual circumstances, active measures

Table 1. Distribution in time of the magnitude of shocks within
a 15-min long seismic event

Event No.	Time/s	Magnitude	Remarks
0	0	0.84	blasting
1	34		
2	103		
3	151		
4	196		
5	223	0.76	
6	234		
7	263	0.89	
8	314	0.24	
9	334	0	
10	367		
11	420		
12	504		
13	578	0.78	
14	682	0.33	
15	689	0.50	
16	798	0.06	
17	750		
18	773		
19	823		
21	871		

can be taken to prevent the occurrence of a rockburst or passive measures to limit the effects of a rockburst in a mine. Active measures should play major role in controlling tectonic rockbursts (Type 2). These are aimed at:

— Minimising the effects of mining on tectonic faults or stress-accumulating strata, and choosing mining methods which take into account the structure of the surrounding rock mass;

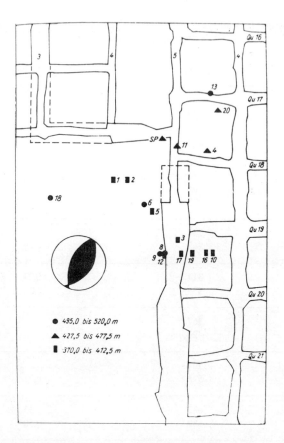

Fig. 18. Localisation results and orientation of the focal plane of the
investigated seismic event (s. Table 1, Behrens *et al.*, 1987).

— Applying measures limiting the subsidence rate such as safety pillars,
unmined bands of the deposit, highly effective filling etc.; and
— Bringing the mechanical state of tectonic fracture zones from the pre-
to the post-failure state, where they can deform without emission of energy.

So seismic methods to characterise rockbursts are an essential prerequisite
for the application of an effective strategy of rockburst control. Seismic methods
are advantageous also because they can be used over long distances and, therefore,
a favourably arranged seismic network can exist for a long time independently
of local mining activities. The measures have no direct influence on the mining
process and yield results which can be computed by modern data-processing
technology. Simultaneously, an analysis of rock behaviour and control of the
effectiveness of measures can be provided.

REFERENCES

Aki, K. 1979. Characterization of barriers on an earthquake fault, *J. Geophys. Res.*, 84, 11, 6140-6148.

Behrens, H.J., E. Scheffler, E. Moog, and G. Rösing. 1987. Seismische Grubenüberwachungssysteme bei der Lösung gebirgsmechanischer Probleme im Kalibergbau der DDR, *Freiberger Forschungshefte*, C 424, pp. 45-53.

Das, S. and K. Aki. 1977. Fault plane with barriers: A versatile earthquake model, *J. Geophys. Res.*, Vol. 82, pp. 5658-5670.

Das, S. and B. V. Kostrov. 1983. Breaking of a single asperity: Rupture process and seismic radiation, *J. Geophys. Res.*, 88, B.5, 4277-4288.

Das, S. and B.V. Kostrov. 1985. An elliptical asperity in shear: Fracture process and seismic radiation, *Geophys. J. R. Astr. Soc.*, vol. 80, pp. 725-742.

Gendzwill, D.J., R.B. Horner and H.S. Hasegawa. 1982. Induced earthquakes at a potash mine near Saskatoon, Canada, *J. Earth Sci.*, vol. 19, pp. 466-475.

Gibowicz, S.J., A. Cichowicz and T. Dybel. 1977. Seismic moment and source size of mining tremors in Upper Silesia, Poland, *Acta Geophys. Polon.*, vol. 25, pp. 201-208.

Gibowicz, S.J., A. Bober, A. Cichowicz, Z. Droste and J. Hordejuk. 1979. Source study of the Lubin, Poland tremor of 24 March 1977, *Acta Geophys. Polon.*, vol. 27, pp. 3-38.

Gibowicz, S.J., A. Cichowicz, A. Bober and M. Kazimierczyk. 1980. Calculations of seismic energy of rockbursts in the Lubin copper mine, Poland, *Acta Mont. UGG CSAV*, vol. 55, pp. 15-35.

Hinzen, K.G. 1982. Source parameters of mine tremors in the eastern part of the Ruhr District (West Germany), *J. Geophys.*, vol. 51, pp. 105-112.

Hurtig, E., H. Grosser, P. Knoll and H. Neunhöfer. 1982. Seismologische und geomechanische Untersuchungen des seismischen Ereignisses vom 23.6.1975 im Werragebiet bei Sunna (DDR), *Gerlands Beitr. Geophys.*, Leipzig, 91, 1, 45-81.

Kisslinger, C., J.R. Bowmann and K. Koch. 1982. Determination of focal mechanisms from SV/P-amplitude ratios at small distances, *Phys. of Earth and Planet Int.*, vol. 30, pp. 172-176.

Knoll, P. 1979a. Zum Bruchmechanismus gedrungener Bergfesten bei dynamischer Beanspruchung, *Rock Mech.*, Suppl. 8, pp. 209-226.

Knoll, P. 1979b. Discussion on the 4th Int. Congr. On Rock Mech., Montreux. In: *Proc.*, vol. 4, pp. 406-408. A.A. Balkema Publ., Rotterdam.

Knoll, P. 1981. Gemoechanische Modellvorstellungen zum Mechanismus spröder Brüche des Gebirges in Bergbaugebieten. Diss. B., Bergakademie Freiberg.

Knoll, P. 1981. Die Energiebilanz von Gebirgsschlägen, ein Mittel zur Bestimmung ihres geomechanischen Mechanismus, *Neue Bergbautechnik*, 11: 7, 384-389.

Knoll, P., A. Schwandt and K. Thoma. 1979. Die Bedeutung geologischtektonischer Elemente im Gebirge für den Bergbau, dargestellt am Beispiel des Werra-Kalireviers der DDR. In: *Proc. 5th Symp. on Salt, Hamburg (FRG) 1978: North. Ohio Geol. Soc.*, vol. I, pp. 105-113.

Knoll, P., K. Thoma and E. Hurtig. 1980. Gebirgsschläge und seismische Ereignisse in Bergbaugebieten. *Rock Mech.*, Suppl. 10, pp. 85-102.

Knoll, P., W. Kuhnt, J. Sievers and G. Rösing. 1984. Verfahren zur Ermittlung der Auswirkungen unterirdischer Brüche, WP G 01 V/272 205 8.

Kosyrev, A.A. 1985. Anwendung eines Kolpexes hochgenauer geophysikalischer Methoden für die Kontrolle der natürlichen und technogenen Deformationen hochbeanspruchter Felsmassive. Vortrag zur XII. Wiss. Koordinierungsberatung. RGW-Thema 09.10, Schlottwitz, DDR.

Kuhnt, W. 1985. Spektrale Untersuchung und Stärkebestimmung seismischer Ereignisse in Bergbaugebieten. Diss. A, AdW, Forschungsbereich Geo- und Kosmoswissenschaften Potsdam (unpublished).

Kuhnt, W., H. Grosser, E. Hurtig and P. Knoll. 1984. Focal parameter studies of local seismic events by means of different source models. European Seismological Commission, XIX General Assembly, Moscow, October 1-6, 1984.

36

Kuhnt, W., P. Knoll, H. Grosser and H.J. Behrens. 1987. An attempt to estimate the damage in mines due to seismic events and rockbursts on the basis of determination of spectral source parameters, *Gerlands Beitr. Geophys., Leipzig*, 96, 3/4, 311-319.

Krishnamurthy, R. and K.S. Nagarajan. 1983. Rockbursts in Kolar goldfields. In: *Rock Mechanics, Proc. Indo-German Workshop, Hyderabad, Oct. 1981, New Delhi, 1983*, pp. 125-140.

Madariaga, R. 1976. Dynamics of an Expanding Circular Fault, *Bull. Seism. Soc. Amer*, vol. 66, pp. 639-666.

Rösing, G., H.J. Behrens, W. Kuhnt and W. Fleischer. 1987. Application of simple source models for description of the mechanism of small seismic events in mining regions, *Gerlands Beitr. Geophys., Leipzig*, 96, 2, 162-170.

Salamon, M.D.G. and H. Wagner. 1979. Role of stabilizing pillars in the alleviation of rockburst hazard in deep mines. In: *Proc. 4th Int. Congr. Rock Mech., Montreux*, vol. 2, pp. 561-566.

Shemjakin, E.I. 1987. The mechanics of rockbursts. In: *Fiz. probl. raz. pol. iskop., Novosibirsk.* (In Russian)

Smirnov, V.A., G.M. Gelashvili, S.S. Shatalow and Z.A. Gorbeziani. 1973. Complex geophysical investigations of rockbursts in mines of the Tkibuli deposits. In: *Trudy VNIMI*, vol. LXXXVIII, pp. 222-230, Leningrad. (In Russian)

Smith, R.B., P.L. Winkler, J.G. Anderson and C.H. Scholz. 1974. Source mechanisms of microearthquakes associated with underground mines in eastern Utah, *Bull. Seism. Soc. Amer.*, vol. 64, pp. 1295-1317.

Spottiswoode, A.M. and A. McGarr. 1975. Source parameter of tremors in a deep-level gold mine. *Bull. Seism. Soc. Amer.* vol. 65, pp. 93-112.

Wong, I.G. 1985. Mining-induced earthquakes in the Book Cliffs and eastern Wasatch plateau, Utah, USA, *Int. J. Rock Mech. Min. Sci. Geomech.*, Abstr., 22, 4, 263-270.

3

GEOPHYSICAL EXPRESS METHODS OF ROCKBURST PREDICTION

V.M. Proskuryakov

VNIMI, Leningrad, USSR

A large variety of mining-geological and mining-technical conditions of mineral deposit exploitation, development and deepening of mines, considerable changes of physical-mechanical qualities of the rock massif, predominance of horizontal field stress components over vertical ones and several other factors cause dynamic manifestations of rock pressure such as rockbursts of different intensity (bursts, bumps, micro-bursts and rockbursts). Also, intensive formation of benches and peeling indicate danger of rockbursts in the massif and ore (rock) at different stages of rock pressure development.

VNIMI had developed several new geophysical express methods of prediction of rockburst danger level in coal mining regions and ore deposits for accurate prediction and prevention of the above-mentioned phenomena. These methods make possible the control of mining without disturbing the technological process (vibroseismic method, method of induced electro-magnetic radiation), and also to effect control at the proving stage.

These methods are useful in controlling the efficiency of preventive methods (for prevention of rockbursts). These are camouflet blasting, infusion of water into coal bed and some others. Changing the difficult geomechanical methods to rapid geophysical ones considerably reduces the time duration of diagnosis and is much cheaper too.

Physically, the use of geophysical methods for prediction of rockburst danger level in coal mining regions is based on the existence of simple interconnection of parameters of natural or artificially induced physical fields in the massif with change in state of strain.

The method of induced high frequency acoustic emission (vibroseismic method) is based on direct dependence of intensity of artificial signals, generated during boring of technological shafts, from the existing bearing pressure of the massif.

Brittle destruction of rocks occurs in the process of drilling and is followed by the appearance of high frequency impulses of small amplitude. Besides, dynamic influence of the drilling tool on the hole causes low frequency fading oscillations of large amplitude in the massif. These oscillations, spreading in the massive are summed up at the point of registration and recorded by a receiver installed at the outcrop of the massif.

During this process, the conditions of occurrence of artificial signals generated in the massif in the process of blast-type drilling, change. While drilling in the region of higher stresses, the impulses occurring during brittle destruction will have a higher value of intensity caused by the presence of forces of elastic rock deformation in the process of destruction. Simultaneously, with drilling at the zone of maximum bearing pressure, the intensity of impulses, which appear under the dynamic influence of the boring tool on rocks being destructed, will grow. This happens because of the transmission of a larger part of the burst energy in the massif that occurs with an increase in acoustic rock rigidity in this zone. This method is used in "VOLNA" instruments designed by VNIMI.

The method of acoustic emission is based on recording the impulse of elastic oscillations radiated during the formation of compression joints. The highest level of acoustic emission frequency (the number of impulses per unit of time) is observed during the deformation of rocks beyond the ultimate strength in the descending part of the stress-deformation diagram. Therefore, it is possible to prove the evidence of one of the conditions of rockburst, which is rock deformation beyond the ultimate strength by the frequency of acoustic emission. The other condition of rockburst danger is the unstable condition of the whole system, which includes the destructive element of the massif and surrounding rocks. One sign of it is the appearance of a zone of deformations formed from the prevailing brittle destructions (in the side of the massif). The energy and accoustic emission impulse amplitude following these processes increases accordingly. So the correlation between acoustic emission impulses of different energetic classes (amplitude distribution) is a characteristic feature of instability of rock condition.

Therefore, to assess the degree of rockburst danger of a rock massif region according to natural acoustic emission, it is necessary to use two parameters— intensity and index of amplitude distribution of impulses. Measurements by this method are conducted by instrument SB32, designed in VNIMI.

The method of natural electromagnetic radiation is based on the influence of mechanical loads. This effect is followed by electrical discharge, bridging a gas gap between the sides of joints and by the appearance of impulse electromagnetic radiation in the spectrum of radio waves. Electrical fields by friction and discharges in rocks are caused by deformation followed by a shift of joints, displacements and failures, and leads to consolidation of some blocks and grains between them. Thus, in the preparation process of dynamic phenomena there will be continuous transformation of mechanical energy into electrical in the rock massif, followed by electromagnetic radiation. The growth of brittleness, modulus of elasticity, strength and stresses of the massif and hence degree of rockburst danger contributes to an increase in frequency of electromagnetic radiation impulse. For measuring this, EQ-6M instruments designed in VNIMI are used.

The method of recording induced electromagnetic radiation is based on registration of electromagnetic waves, formed by brittle destruction of the highly stressed coal (rocks) near the moving surface of the drillhole, when drilling in the side of the massif. The impulses of the induced electromagnetic radiation are received within the frequency band of 10-150 kHz. Magnetic component of the field is recorded on one of the band frequencies, which provides a 10 dB signal-to-noise ratio. A magnetic antenna and impulse analyser, tuned for resonance, are used for registration. The antenna is placed 1-1.5 metres from the mouth of the hole being drilled. The number of above-edge amplitude signals are registered in similar intervals of movement of the blasthole surface.

Charts are then made of dependence of number of electromagnetic radiation impulses upon distance to the massif outcrop, according to which one can evaluate the danger of rockbursts. The method of electromagnetic probing is based on the dependence of specific electrical resistance of rocks upon concentration of stresses in the zone of bearing pressure. Decreased resistance with increased bearing pressure on the side of the massif is characteristic of ores, coal and rocks. Rock deformation near exploration openings is followed by the formation of an anomalous change in the rock-resistance region, while in the undisturbed massif, the resistance level is different. The newly formed region consists of two zones, differing by the stress-deformation condition of rocks: the first zone consists of dislocated rocks with residual strength from the perimeter of the exploration opening; the second zone consists of rocks capable of accumulation of elastic compression energy and of release in the form of dynamic phenomena. Rocks in the second zone are influenced by higher bearing pressure. Resistance of rocks in the second zone decreases on approaching maximum bearing pressure and increases above maximum, gaining maximum level at a distance approximately equal to the width of zone stressed to the limit of rocks beyond maximum. The degree of rockburst danger is evaluated by the ratio of maximum value of resistance in the second zone.

The method of electromagnetic probings is based on inductive excitation of a high-frequency electromagnetic field in the massif. This method does not demand contact with coal (rocks) at the exploration opening. For probing, we use KAE-1 or EQ-6M, designed in VNIMI.

All these express methods of rockburst prediction have been successfully tested and are widely used in coal and ore deposits in the USSR.

ROCKBURST EXPERIENCES AND RESEARCHES IN KOLAR GOLDFIELDS

4

ROCKBURST HAZARD AND ITS ALLEVIATION IN KOLAR GOLD MINES—A REVIEW

I.M. Aga, P.A.K. Shettigar and R. Krishnamurthy

Bharat Gold Mines Ltd., Kolar, Karnataka, India

INTRODUCTION

Kolar Goldfields is situated at 12°57′ N and 78°16′ E in the south-east corner of Karnataka State near Bangalore city in India and lies at an altitude of 900 m above mean sea level. There is indication that some of the ancient workings are more than 1,000 years old. However, the modern phase of mining started in 1880 and has continued ever since. Approximately 49 million tonnes of ore have been mined, yielding 793 tonnes of gold.

The problem of ground control and rockbursts associated with hard rock has been present in Kolar gold mines since the beginning of the century. Rockbursts have occurred at all depths under different mining conditions. However, the problem became more serious as mining reached greater depths. One of the mines has reached a depth of over 3.2 km. Rockbursts have caused large-scale damage to underground workings including loss of shafts, travelling and haulage roadways, pumping and winding installations. Surface buildings have also been extensively damaged. Fatalities are generally associated with rockbursts. Valuable proved ore reserves have been lost forever. However, the problem of rockbursts has been considerably reduced by the introduction of better mining methods, based on studies of rock mechanics.

GEOLOGY AND STOPING METHODS

The geology and mining practise of KGF have been discussed in detail by Pryor

[1] and Taylor [2]. The mines of KGF occupy a strike length of approximately 8 km and the deposits occur in a belt of hornblende schist of lower Dharwar age, which is surrounded by granite and gneisses. The lodes dip towards the west at 40° to 45° near the surface, gradually changing to nearly vertical at depth. Of the many quartz lodes explored on the field, only two lodes, Champion and Oriental, are of great economic importance. The width of the lodes varies from 1 m to 6 m. Productive workings on these lodes have been interrupted by a series of faults, pegmatites and dykes. The Champion lode has been extensively worked in all three mines—the Mysore, Champion Reef and Nundydroog—which are operating at present, while the Oriental lode is being mined on a large scale only in Nundydroog mine in the northern part due to economic reasons. In Champion Reef mine, the main workings are on the Champion lode at depths of over 2,800 m.

In KGF development work for extraction of lode is generally done as per standard metal mining practise. Shafts at depth are usually supported by brick or reinforced concrete lining and the levels, where necessary, are supported by steel sets, lagged and packed. Extraction of lode is done mainly by (i) underhand stoping with granite packing as the medium of support, (ii) stope driving with concrete as support, and (iii) flatback stoping with sand-fill as support. The rill system of stoping which was practised at depth in Champion Reef Mine earlier has been discontinued and replaced by stope driving.

ROCKBURSTS

According to early records [3], the first rockburst reportedly occurred in a stope below 9960 ft in Oorgaum mine, now a part of Champion Reef mine, in 1898. Rockbursts were classified in the early days as 'Air-blasts' and 'Quakes' depending on their intensity and area of damage.

At shallow depths, these problems were not critical except while mining shaft pillars and in exceptional cases ore shoots which were highly stressed due to juxtaposition of faults. However, they became serious towards the 1930s as mining reached greater depths, particularly when the ore body to be mined was associated with faults, dykes, and pegmatites, all involving a plane of weakness. Large rich ore shoots have been completely damaged and rendered unproductive as a result of severe rockbursts and, in these cases, the first rockburst tremor was followed by a series of tremors over a period of several days. Surface buildings within 2-3 km from the epicentral region were damaged. The intensity of some major rockbursts was in the range of 5.0 to 6.0 on the Richter scale and they were recorded by seismographs located as far as 760 km away from KGF. Figure 1 shows the typical damages to steel-set roadways as a result of rockbursts. A Weichert seismograph was installed on the surface in 1912 and rockburst tremors of magnitude ≥ 1 mm style displacement recorded by

Fig. 1.

the seismograph are shown in Fig. 2.

Details of typical major rockbursts resulting in severe damages to underground workings are given in the Appendix. Of these, rockbursts that occurred in Glen ore shoot on 27th November 1962, and Northern folds on 25th December 1966 in Champion Reef mine and in the west reef workings of Nundydroog mine on 27th November 1971 were the most severe [4].

In Glen ore shoot of Champion Reef mine, which extended from level

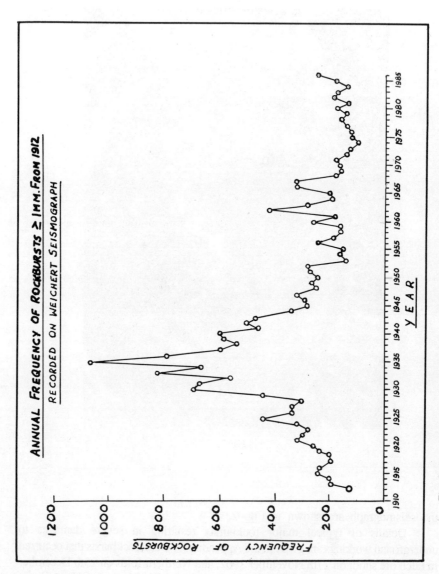

Fig. 2. Annual frequency of rockburst—IMM from 1912 recorded on Weichert seismograph.

84 to level 110, where rill system stoping with dry granite support in a 'V' sequence was practised, as many as 98 rockbursts resulting in damage to workings occurred during the period 1942-62. Rockburst occurrences in the area may generally be classified in two groups: one, where damage was confined to the stope and stope abutments, and the other, to footwall drives and complementary crosscuts east of a major fault, known as Mysore northern fault, striking NNW converging on the ore shoot on the northern wing. In this area, the frequency of occurrence of minor and medium rockbursts was one in every 58 days and in the major area, one in 246 days.

Of the above rockbursts, the one that occurred on 27th November 1962 between levels 85 and 107 of Glen ore shoot covering an area of 300 m on strike and 450 m vertically was the most severe. It was the first of its kind in the history of mining in Kolar gold fields. Surface buildings in the Champion Reef mine area were also damaged. The first major tremor was followed by a series of tremors, some of equal intensity, which continued for several days. A total of 59 tremors were recorded on the Weichert seismograph during the following 24 hours. Extensive damage occurred to the 'V' 'Sawtooth' abutments. Heavy damage also occurred in the crosscuts lying immediately opposite the stope face and in the abutments. Heavy falls of ground occurred mostly in the unsetted portions of shafts, crosscuts, complementary crosscuts and footwall drives remote from the stope face. Approaches opposite the stoped ground were affected least (Fig. 3).

In the Northern folds of Champion Reef mine where the reef formation is 'Z'-shaped in plan view and extends from level 73 to level 113, as many as 69 rockbursts occurred from 1943 to 1966, resulting in damage to workings.

Biddick's shaft, now known as the Sub-Auxiliary Shaft, which is the main access to the area on the south side, lies approx. 300 ft. in the footwall of the reef and has been damaged on 14 occasions. On 55 occasions, the stoping area has been affected. The frequency of rockbursts of minor to medium intensity was one in 33 days and that of major intensity, one in 421 days. Rill system stoping was done with dry granite support adhering to a 'V' sequence in each limb with a lag and lead between stoping sequences in the limbs. Damage was mostly confined to abutments in this stoping area also, just ahead of stope faces, as was seen in Glen ore shoot.

Of the rockbursts that have occurred in Northern folds, the one on 25th December 1966 affecting the ground below level 97 was the most severe. The first major tremor was followed by a series of tremors with as many as 65 occurring during the week ending 31st December 1966. Damage in the stoping area was confined mainly to the abutments, the reef drives being severely damaged. Falls of ground occurred in the unsetted portion of drives and crosscuts. The damage was very severe below level 100. The shaft lining of the sub-auxiliary shaft, the main access to the stoping area on the south side, suffered serious damage between levels 97 and 98 (Fig. 4).

48

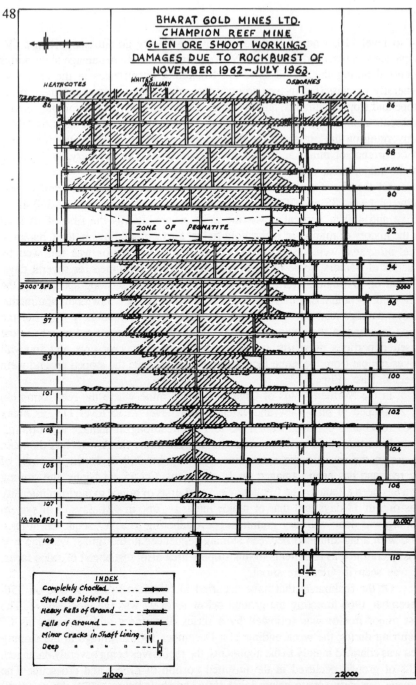

Fig. 3. Champion Reef Mine Glen ore shoot workings damaged due to
rockburst of November 1962.

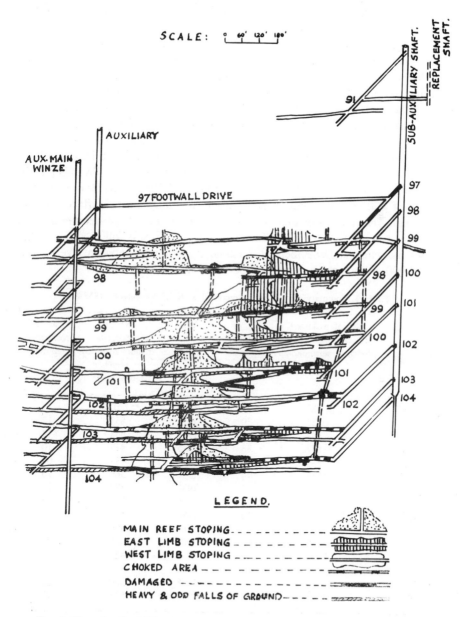

SCALE: 0 60' 120' 180'

Fig. 4. Champion Reef Mine. Part stepped plan showing damages due to rockburst on 25.12.1966.

It is interesting to note that the bottom sections of Champion Reef Mine had been flooded due to unprecedented rains 3 months before the burst. A large number of rockburst tremors were recorded in the field particularly during those three months.

In the west reef stoping to level 62 in Nundydroog mine, where extraction is done by the flatback system with hydraulic sand-fill as support, as many as 94 rockbursts occurred during the period 1957-72, nearly 41% occurring during the period when the stope was in the process of holing through to the level above. The damage due to rockbursts in the west reef workings differed from that experienced in the workings of Champion Reef mine. Geologically, this lode is quite different from the main lode of Champion Reef mine and commonly intersected by NE-striking faults and by dykes striking E-W.

Of the rockbursts, the most severe one occurred on 21st November 1971, resulting in damages to the stoping area from level 3,650 to 6,200 over a strike length of 400 m with significant damage to buildings on the surface for over 3 sq km. The first major tremor was followed by a series of tremors at short intervals and bursting continued for several days. This was followed by another major rockburst in December 1972 damaging workings between levels 5,000 and 5,900 over a strike length of 300 m.

Table 1. Summary of rockbursts recorded at KGF observatory 1957-86

Year	No. of rockburst tremors recorded on Weichert seismograph	No. of rockbursts reported	Year	No of rockbursts tremors recorded on Weichert seismograph	No. of rockbursts reported
1957	387	23	1972	**1202	14
1958	537	30	1973	852	12
1959	549	29	1974	643	14
1960	877	33	1975	419	10
1961	589	47	1976	434	09
1962	*851	44	1977	366	10
1963	*632	43	1978	479	14
1964	478	21	1979	539	18
1965	748	12	1980	475	11
1966	*1096	04	1981	494	21
1967	1109	09	1982	485	23
1968	1157	05	1983	*422	20
1969	1480	02	1984	364	26
1970	762	03	1985	359	20
1971	**1155	10	1986	577	12

*Major area rockbursts in Champion Reef mine.
**Major area rockbursts in Nundydroog mine.

ROCK MECHANICS RESEARCH IN KOLAR GOLD FIELDS

Investigations of the problem of ground control and rockbursts in KGF were undertaken in the past by a few individuals working in the field but in 1926 and 1955 special committees were set up for this purpose. The report of the 1955 committee [3] contained comprehensive recommendations for safe mining practise. The incidence and severity of rockbursts and accidents to workmen decreased as a result of adopting the committee's recommendations in the form of standard mining practises. However, systematic investigation of these problems was introduced in 1955 when a Rockburst Research Unit was formed and has continued ever since. Work until 1972 was done in collaboration with the University of Newcastle-upon-Tyne, U.K., under the guidance of late Prof. E.L.J. Potts.

The investigations carried out consisted of statistical analysis of available data, laboratory tests on the strength and elastic properties of rocks and field measurements on rock-mass movement occurring in and arround underground excavations.

Statistical analysis of the records of rockbursts revealed the following [5, 6]:

1) Only a small percentage of the rockbursts recorded on a Weichert seismograph located on the surface were traceable underground. This is so even for bursts recorded by the recently installed surface and underground seismic network. Rockbursts recorded by the Weichert seismograph and those reported from mines resulting in damage are shown in Table 1.

2) Frequency of reported rockbursts is minimum on Sundays and maximum on Fridays.

3) A significant peak in occurrence of rockbursts is seen during the time of stope blasting or soon thereafter.

4) A specific correlation between rainfall and rockbursts has been noted.

5) Frequency and severity of rockbursts are not directly related to depth. A large number of rockbursts of medium and major intensity have also occurred at shallow depths. The important factors for the causes of rockbursts are the physical and elastic properties of rocks, *in situ* stress, size and shape of excavations, and inhomogeneity of rock such as the existence of faults, pegmatites, dykes and calcite stringers, all involving a plane of weakness either in themselves or at contact.

6) The major area rockbursts resulting in very severe damage to underground workings and surface buildings have occurred in the field at approx. 10-year intervals.

7) In the case of major area rockbursts affecting large ore shoots, recurrence of spitting and arching in the footwall drives and crosscuts was experienced over a period of two to three weeks prior to rockbursts.

LABORATORY INVESTIGATIONS

Laboratory investigations conducted since 1957 have provided valuable information on the strength and elastic properties of KGF rocks [7, 8].

Kolar schist has an average uniaxial compressive strength and Young's modulus of elasticity of 3,000 kg/sq cm and 7.9×10^5 kg/sq cm respectively, Poisson's ratio being 0.2. The reef quartz has a high compressive strength of the order of 4,200 kg/sq. cm. The results of tests justify defining Kolar schist as a transversely isotropic medium. The violence at rupture of KGF rocks is 2 to 3 times higher than in South African quartzite.

UNDERGROUND INVESTIGATIONS

In Kolar Goldfields high inherent stresses exist in virgin rock that cannot be accounted for by superincumbent weight alone. In some cases, lateral stresses are higher than the vertical stress field and the ratio of lateral to vertical stress field varies from 1.6 to 4.0. Stress measurements were done using the stress-relief technique, strain being measured by an extensometer and sonometer in vogue in the early days.

However, *in-situ* stress measurements done in the 1980's using hydrofracturing techniques underground have shown that the vertical stress field varies from 35% to 70% of the overburden pressure, and the principal horizontal stresses are independent of the overburden pressure and are probably tectonic phenomena [9].

Large-scale rock mechanics instrumentations were carried out in and around stopes which had been worked under different mining conditions and subjected to severe rockbursts:

i) Rill stopes supported by dry granite packwalls at depths to 2,758 m in Champion Reef mine, and

ii) Flatback sand-filled stopes at depths to 1,750 m in the Oriental lode of Nundydroog mine.

Based on these investigations, new mining methods consistent with safety have been developed.

Champion Reef Mine

In and around rill stopes with dry granite support at depth of 2,315 m-2,758 m [6, 8, 10, 11] (Figs. 2, 3).

1) Displacement pattern parallel to the reef plane indicated an extension zone in advance of the face and compression zone behind the face bearing a strong similarity to the pattern observed above excavations in horizontal seams.

This was so even in sand-filled flat back stopes at depths to 1,750 m in Nundydroog mine.

2) At normal to the reef plane, measurements showed an erratic compressional movement in the foot- and hanging walls of reef in advance of the face line, indicative of an unstable equilibrium there. Immediately behind the face and opposite the stoped-out ground, the wall rock dilated. Further behind, there was recompression.

3) Closure between stope walls behind the stope face was generally erratic and the maximum measured closure did not exceed 12-15% of stoping width though the voidage in the support was 34%. In some cases, wall closure continued as far as 50 m behind the face indicating that full consolidation of support was not attained for a considerable distance behind the face. Thus a state of instability existed over a considerable distance behind the face.

Wall closure in the sand-filled flatback stopes was more uniform, the maximum closure being of the order of 8 to 10% of stoping width in Nundydroog mine.

4) At the reef plane, a "downward" movement of the hanging wall relative to the footwall was seen at places behind the face. In advance of the face, the footwall rock generally "slumped" relative to that of the hanging wall. Slumping of the footwall rock has been a marked feature during periods of rockbursts.

This was also observed in the sand-filled flatback stopes in Nundydroog mine.

5) Stress measurements in advances of stope faces indicated the existence of very high front abutment stresses at distances varying from 1.2 to 1.8 metres in advance of the face. The maximum load measured in the granite packwall of a rill stope at a distance of 3.5 m behind the face was much lower than expected.

In the front abutment of sand-filled flatback stopes in Nundydroog mine, stress increases of the order of 600-1000 kg/cm^2 were measured at 0.5 to 1.5 m in advance of the face but the load in the sand-fills registered low values.

It was concluded that the solid granite supports used in Champion Reef Mine at depth and the slow rate of extraction (both of which have a major and adverse effect on mining economics) did nothing to assist in providing controlled relief to the wall rock nor to minimise rockbursts. On the contrary, they assisted in building high stresses on the solid abutments in advance of the faces and created a situation that went beyond control. Hence, it was found absolutely necessary to change the rill system of mining to a new system which would result in better ground control and recommendations were as follows:

a) Adopt a technique of geometrically designed sequence which permits a considerably increased rate of advance to be achieved and maintained; limit the area to be exposed after each machine cycle to the minimum and adopt a quick support system;

b) Adopt a support system which yields and provides relief to the rockmass but is sufficiently strong to support the immediate walls, thus deliberately assist the main body of the rock mass to attain static and dynamic equilibrium quickly.

A system of ''stope drive'' with concrete supports was practised for stoping rockburst-prone ore shoots and pillars at shallow depths, and generally satisfied the above conditions.

Hence, the Northern fold area below level 97, severely damaged by rockbursts on 25th Dec. 1966, was reopened and extraction commenced by ''stope drive'' system with three longwall faces, i.e., between levels 100 and 103, 103 and 105, and 105 and 109 (Fig. 5). Extraction was first restricted to the east limb but has now been extended to the main reef between the same levels with a lag and lead between longwall faces. Since the commencement of stoping in 1971, to date only 12 rockbursts have been reported from this area versus 37 reported for a period of 8 years between 1958-66 when stoping was being done by rill system.

Similarly, the same stope drive system was adopted to extract the area in Glen ore shoot affected by rockbursts on 27th November 1962. Approaches to the ore shoot were reopened, rill promontories were shaped and a longwall system of stoping was adopted between levels 98 and 103 on the south wing followed by the north wing. The sequence of stoping is maintained in such a way that stope faces below the level always lie one stope length in advance of the stope face immediately above the level. Considerable stoping has already been carried out by the adoption of this method, with very few ground control problems (Fig. 6).

The measured closures in stopes with concrete support have proven to be more regular and uniform with advance of stope face and stability behind the face is reached much earlier compared to those in stopes supported by granite walling. The average total closure between walls in the stopes supported by concrete is approx. 8 to 10% of stoping width over a face advance of 36.5 m. Closure as much as 12 to 15% of stoping width is reached over a distance of approx. 50 m behind the face.

It is interesting to note that during the 10-year period prior to the major rockbursts on 27th Nov. 1962, 58 rockbursts of major to minor intensity occurred resulting in damage to working and, in most cases, the damage was very severe, while for a period of 15 years since the commencement of longwall stoping in 1971 by stope drive system, only 14 rockbursts have occurred resulting in medium and minor damage.

Nundydroog Mine

Hitherto, the general impression was that sand-fill being a yielding support and also homogeneous should provide adequate relaxation to general strata movement of the rock mass compared to rigid granite supports, and that stability is reached much sooner in sand-filled back stopes than in granite-filled stopes under similar

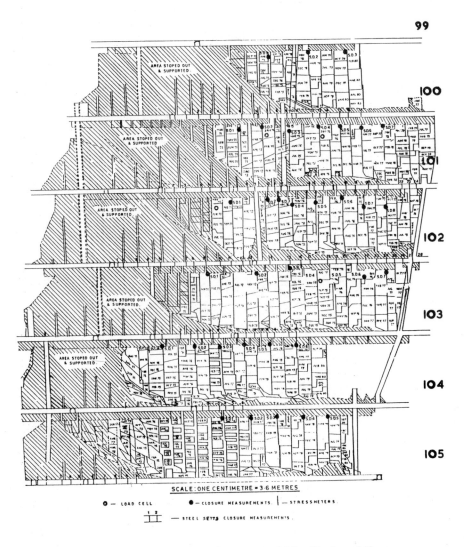

SCALE: ONE CENTIMETRE = 3·6 METRES

○ — LOAD CELL   ● — CLOSURE MEASUREMENTS. | — STRESSMETERS.

— STEEL SETTS CLOSURE MEASUREMENTS.

Fig. 5. Bharat Gold Mines Ltd. Champion Reef Mine Northern folds area.
East limb of fold.

56

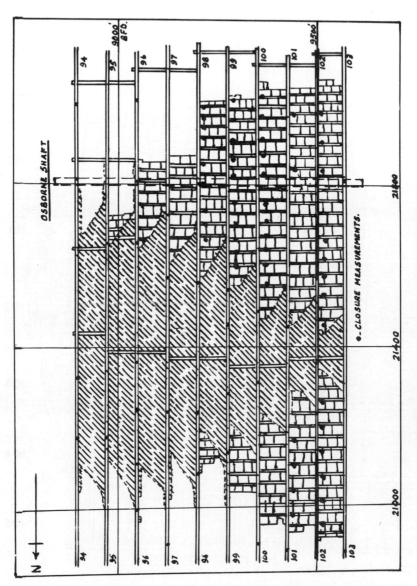

Fig. 6. Stope drive system—Glen ore shoot Champion Reef Mine.

mining conditions. In this case, the wall rocks and reef should become sufficiently fractured to an appreciable distance ahead of the face, shifting the front abutment peak sufficiently forward but, contrarily, the front abutment peak is found to lie very close to the face. Also, the closure measured in sand-fill supports to date has only been 8 to 10% of stoping width (to be treated with some reservations). It would appear, therefore, that either the sand-fill placed in the stope consolidates quickly and assists in transmitting load or the immediate wall rocks, due to their nature, yield only to a limited extent and stand like rigid beams supporting the load and inhibiting closure between walls and general relaxation of strata around the stoping excavation. However, it is considered that though sand-fill consolidates quickly, it should be capable of yielding to such an extent as to permit adequate wall closure, if called upon. Hence, the latter situation may continue until a 'critical span' is reached resulting in an area resettlement or rockburst. In the light of knowledge gained from investigations conducted so far, and past experience the following stoping practises have been adopted to minimise ground control problems:

1) Flatback stoping system with short strike lengths, the stope face advancing as one single face or in steps and the sequence of stoping being maintained in such a way that the stope face below the level is in advance of the stope above the level by one stope length. Stope support is by hydraulic sand-fill as practised hitherto.

2) Wherever the ore body is intersected by dykes and faults, stoping has commenced from the dyke or fault position and moves away from them instead of advancing towards them.

3) Whenever a stope advances to within 9 m from the level, holing is effected at 6 m length at a time by either breasting or bottom stoping.

4) Faster rate or extraction is emphasised.

CLOSURE BETWEEN STOPE WALLS

Analysis of closure measurements between stope walls carried out in selected areas has provided valuable information on the consolidation of fill and general strata movement etc. and the following are considered of great interest [12, 13].

1) Measured closure consists of somewhat gradual closure due to mining and large sudden closure due to rockbursts.

2) Rate of closure at a point was usually high near the face and gradually decreased as the face moved away from it.

3) Magnitude of maximum closure measured varied from one point to another and also depended on mining methods and support system adopted, stress conditions and behaviour of wall rocks.

4) An increase in rate of closure was associated with machining operations carried out in the stope and rockbursts.

Analysis of wall closure measurements in stopes have indicated a correlation somewhat between rockbursts and rate of closure. In a few instances a significant increase has been noted in rate of closure varying from 2 to 16 times the normal rate for a few days prior to a rockburst, while in others there has been no such indication, as shown in Fig. 7. Statistical analysis of results also confirmed the existence of a significant difference between periods and between days before the occurrence of some rockbursts. The sudden large closure which occurred as a result of a burst was followed by a decrease in rate of closure and the normal periodic rate was reached within a period varying from three days to two weeks after the burst.

As closure measurements are found to be very useful in assessing area stability of underground workings, it has now become a standard practise to introduce these measurements in areas vulnerable to rockbursts. A statistical analysis of the results is carried out and a control chart showing 'warning' and 'action' limits is maintained.

Warning limit .. X ± 2a

Action limit .. X ± 3a

where 'X' is the grand average for the preceding three months and 'a' the standard deviation.

Any measurement exceeding the 'action limit' is examined with great care and men withdrawn, if found necessary. Though these measurements have proved very useful in assessing day-to-day stability of stopes, there are limitations with respect to prediction.

MATHEMATICAL MODELS

Application of mathematical models in computing mining-induced stresses and elastic displacements around stoping excavations in hard rock is well known and widely accepted [14]. Of the many models, the following were considered for KGF conditions:

    i) Homogeneous transversely isotropic model extended by Salamon for a vertical reef;

    ii) Homogeneous laminated frictionless elastic model by Salamon.

In (i), the reef is assumed to be infinite and of transversely isotropic material, the thickness of the extracted reef being infinitesimal compared with length, depth and other dimensions. Complete closure is expected to occur. This model was used in KGF to calculate the elastic displacements and strains parallel and perpendicular to the reef with some degree of success. Two cases were considered for the homogeneous laminated frictionless elastic model:

    a) An incompressible reef with a trapezoidal convergence profile in the excavation;

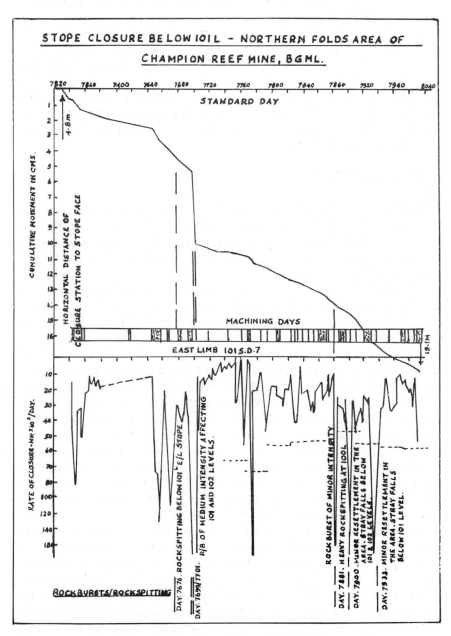

Fig. 7. Stope closure below 101 L.—Northern folds area of
Champion Reef Mine, B.G.M.L.

b) A convergence profile in the excavation based on underground observations.

Mathematical and graphic solutions were obtained. The main limitation of the model is that it cannot be used to calculate movements parallel to the reef plane.

In all cases where the models were used, it was generally noticed that calculated displacements compared favourably (within 15%) with those measured for points remote from the excavations. For points very close to the excavations the differences were large, which might possibly be due to the existence of fractured rock embracing the excavation, which does not accord with the elastic constants used in the models.

SEISMIC INVESTIGATION

Monitoring rockburst tremors to locate their foci and based on seismic activity and anomaly in closure measurements, obtaining timely warning of a impending danger is a great step forward in minimising the hazard. A Weichert seismograph, installed in 1912, provided limited information on some rockbursts and locating their foci was not possible.

Under the Science and Technology Scheme of Bharat Gold Mines Limited, a sophisticated seismic network covering a mining area of 6 km × 3 km was established in KGF during 1978 in collaboration with Bhabha Atomic Research Centre, which has the technical know-how for such instrumentation [15].

A multi-channel seismic network consisting of 14 geophones (7 surface and 7 underground) was established to cover the mining area of 6 km × 3 km. The main objective of the network has been to record rockburst data, delineate zones of high seismic activity, develop velocity models for various regions utilising accurately known coordinates of blasting and improve the accuracy of the computed foci from known rockburst data. The location of surface sensors is shown in Fig. 8.

The signals picked up by the sensors are directly transmitted through a 4-core cable using a carrier frequency of 540 Hz, and recorded on a 24-channel analogue tape recorder at a speed of 15 mm/sec. The recorded analogue tape is replayed onto a fast mingograph recorder to obtain a hard copy of the seismogram for further processing. One of the selected channels of the network is continuously monitored on a helical recorder. Figure 9 shows the seismic recording instruments located on the surface. The frequency coverage reaches 180 cycles and the accuracy of computed foci of rockburst is ± 20 metres. Between Sept. 1978 and 31st December 1986 as many as 9579 rockburst tremors of such intensities as could be recorded on five or more geophones in the network were recorded versus 156 rockbursts reported from the mines resulting in damage to underground workings. Regional velocities have been computed for different areas in the mines, assuming the rock mass to be a homogeneous 'isotropic' medium and also an 'anisotropic' medium. Wherever there was any appreciable

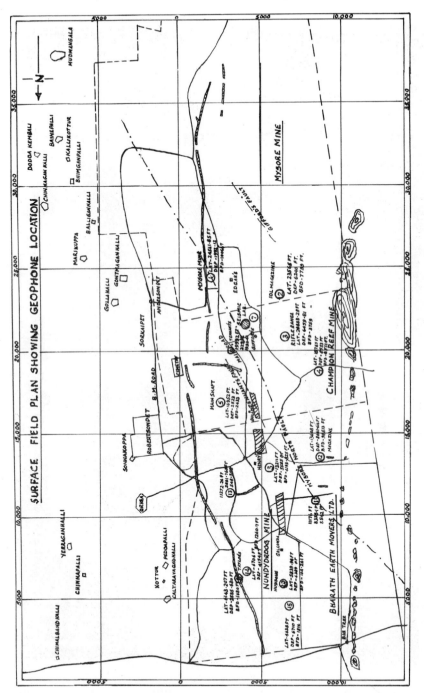

Fig. 8. Surface field plan showing geophone location.

62

Fig. 9. A view of the seismic recording instrumentation system.

difference in the co-ordinates between the computed foci and the location of damage at a particular region, improvements were made by trial and error to obtain a better fit and the most suitable velocities considered for computation of foci, in the event of subsequent rockburst. The computed foci of rockbursts recorded from the bottom section of Champion Reef mine are shown in Fig. 10.

Increase in seismic activity or any anomaly in its trend observed in an area is verified by field observations, including wall closure measurements, and appropriate action is taken immediately. On many occasions work has been suspended and men withdrawn. A mini-computer and a micro-processor are also being used to process the large volume of data recorded by the surface network (Fig. 11).

Seismic monitoring of rockbursts along with closure measurements around stoping regions have been helpful in assessing the stability of stoping regions. The foci of the majority of rockbursts cluster around the current stoping regions, while the remainder are scattered in virgin ground and in old stoped ground containing pillars and remnants.

MICRO-SEISMIC INVESTIGATION

A micro-seismic network of 100 m × 100 m, dimension was established in 1983 between levels 98 and 103 of Osborne shaft area in Champion Reef mine to study micro-seismic activity in the region with a view to forecasting the occurrences of even minor rockbursts [16].

The underground network consists of 10 high-frequency geophones/accelerometers and the micro-seismic signals picked up by the sensors are telemetred through a 4-core cable and fed to a micro-processor through an interface unit located in the surface laboratory. The signals pass through control logic and delay processor of the event threshold detector. The onset of signals of different channels is detected and time delays in a digital form are fed to the micro-processor, which processes the data and prints out the required information on hypocentre parameters. An event counter is also included in the network to count the micro-seismic events in each channel and displayed for a set time to ascertain the rate of micro-seismic activity in the region. The frequency band width covered at present is 100 to 1,000 cycles but is being extended to higher frequencies.

The data collected thus far has indicated in many cases an increase in micro-seismic activity followed by a fall before an event. There are occasions when the event has occurred with a rise in micro-seismic activity. These aspects are being studied in detail to assess event patterns.

CONCLUSIONS

From the investigations conducted to date, it has been possible to obtain, very useful information on stress/strain distribution around workings induced by mining

64

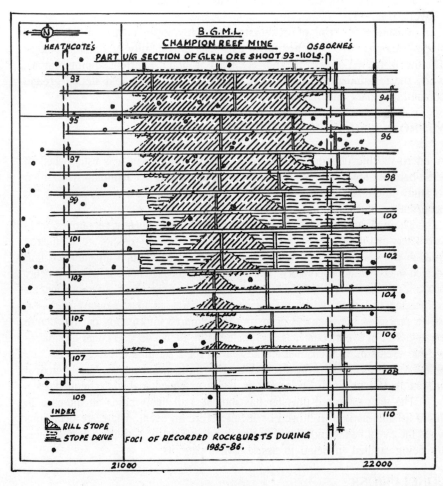

Fig. 10. Champion Reef Mine—FOCI of recorded rockbursts
during 1985–86

Fig. 11. Mini-computer and associated instrumentation array in the laboratory

and strata displacement characteristics of rockbursts. This has greatly assisted planning for improved mining methods in areas vulnerable to rockbursts and thus has reduced the frequency and intensity of bursts.

Seismic monitoring of rockbursts has proved a valuable tool in assessing day-to-day safety of mine workings and on many occasions work is either suspended or resumed depending on the seismic activity recorded from rockburst-prone areas.

Micro-seismic investigations have given positive information in some instances for prediction of bursts.

However, in spite of gathering tremendous information on the phenomena of rockbursts and ground behaviour from investigations carried out so far and planning improved mining methods consistent with safety, rockbursts have continued to occur in mine workings. Though their frequency and intensity are considerably reduced, they continue to elude solution.

ACKNOWLEDGEMENT

The authors are grateful to late Prof. E.L.J. Potts for his inspiration and guidance in conducting the investigations to 1971; the authorities of Bhaba Atomic Research Centre, Bombay, without whose help the seismic and micro-seismic investigations would not have been possible; and the Department of Mines, Ministry of Steel and Mines, Government of India for encouragement and funds for this undertaking.

APPENDIX

Details of Typical Rockburst Areas

Example 1—Mysore Mine and Champion Reef Mine

On 23rd January 1952, a major area rockburst occurred at 0704 hours causing damage underground from just below the surface to the 16th level, a vertical depth of 463 metres in Glen's section of Champion Reef mine and to the 27th level, 614 metres vertical depth in Gilbert's and Tennant's sections of Mysore mine, and for a distance of approximately 762 metres along the strike. The first tremor was followed by as many as 71 tremors during the next 48 hours. The intensity of tremors was so severe that some were recorded at the observatories of Colaba, Bombay, and Kodaikanal. The zone of severe damage was along and parallel to the main fault systems of Champion lode. Many surface buildings along and parallel to this fault zone were severely damaged, the type of damage being somewhat similar to that caused by an earthquake.

Example 2—Champion Reef Mine

A major rockburst occurred in Biddick's shaft section between levels 95 and 102, a vertical distance of 213 metres, at 0119 hours on 11th February

1956. The rockburst was centred on the ground south of the stope face between the 93rd and 96th levels. Shaft crosscuts, junctions of reef drives and crosscuts and steel setted reef drives including stopes between these levels were very severely damaged. The damage was comparatively less in levels between 97 and 103, however.

Damage to the Biddick shaft was very extensive. The entire concrete lining along the east wall between and inclusive of byatts between the 94th and 98th station was completely blown out or very severely damaged. All the shaft equipment such as road guides, byatts, keps and station equipment was either missing or damaged beyond repair from the 95th to the 103rd station.

Example 3—Mysore Mine

A series of major rockbursts occurred on the night of 25th May 1960 affecting a large area in Edgar's middle section, 43rd ore shoot and Edgar's shaft pillar area in Mysore mine. The damage was widespread and extended 600 metres along the strike from Gilbert's shaft to a position approx. 60 metres north of North Reclamation winze and 295 metres along the dip extending from the 15th level Tennant's to the 45 leavel Edgar's. Five shafts, namely, Edgar's, Gilbart's, Tennant's, Hancock's and 32 No. 5 shaft, were affected. The whole area was rendered unproductive for nearly three months.

Many surface buildings were extensively damaged as a result of the rockbursts.

A detailed examination of the affected areas indicated that the rockbursts probably occurred due to some movement along the Mysore north fault.

Example 4—Champion Reef Mine

At 0220 hours on 27th November 1962, a major area rockburst occurred resulting in very severe damage to the whole of the stoping areas on the northern and southern wings of Glen ore shoot between the 85th and 107th levels, a vertical span of 550 metres and to the area known as the Southern Ore Body, with very extensive falls of ground in the Heathcote and Osborne shaft crosscuts, footwall drives and their complementary crosscuts in Champion Reef mine. The first major tremor was followed by a series of tremors and as many as 59 tremors were recorded during the next 24 hours. Abnormal bursting continued for nearly two weeks.

Also a series of major rockbursts occurred between the 16th and 23rd July 1963 resulting in very severe damage, particularly to the southern wing of Glen ore shoot between the 86th and 111th levels. This disturbance was recorded as a slight tremor at the Madras observatory. The whole of Glen ore shoot and the Southern Ore Body were put out of production as a result of these rockbursts.

There was also widespread damage to surface buildings due to these rockbursts.

Example 5—Champion Reef Mine

A series of major rockbursts occurred commencing from 0807 hours on Sunday, 25th December 1966 damaging the Northern folds area below level 97, Auxiliary main winze and Sub-auxiliary shaft at levels 97/98. Reef drives on the main reef, east and west limbs, particularly between the crosscuts joining the main reef and west reef, near the apex and Sub-auxiliary shaft crosscuts, up to level 100 including Sub-auxiliary shaft lining between levels 97 and 98 and the 97th level plat were severely damaged. There were falls of ground in the unsetted portion of drives and crosscuts. Abnormal bursting continued for several days.

Example 6—Nundydroog Mine

A series of major rockbursts occurred in Nundydroog mine commencing from 0210 hours on 27th November 1971 resulting in very severe damage to the west reef workings between levels 3650 and 62 between N14 and N27 area and also to surface buildings in the mining area and BEML area. The reef drives were severely affected over a distance of 45 metres in the 3650, 3800, 3900 and 4000, N17-20 area. Heavy falls occurred in the 4800 N18, 5400, 5900, 6000 and 6200 west reef crosscuts. Most of the underground damage was concentrated between N17 dolerite and N23 porphyry dykes. Damage to surface buildings was significant over an area of 3 sq km.

Example 7—Nundydroog Mine

On 28th December 1972 a major rockburst occurred resulting in severe damage to underground workings between levels 50 and 59, N20-30 area on the west reef and minor damage to surface structures.

Example 8—Champion Reef Mine

A major rockburst occurred on 1st and 2nd October 1983 damaging the stoping area between levels 97 and 108 on the east limb and main reef, northern folds of Champion Reef mine. Damage on the east limb stopes between levels 106 and 109 and stopes on the main reef between levels 100 and 103 was heavy. All the damage was confined to areas behind the stope faces. The first tremor was followed by a series of tremors and as many as 32 tremors having their foci below level 97 on the east limb were picked up by the seismic network during the week.

Example 9—Champion Reef Mine

A major rockburst occurred on 19th May 1985 resulting in minor to medium damage to stoping areas from levels 97 to 101 Northern fold and major damage to the Sub-auxiliary shaft from levels 97 to 101 and level 97 footwall haulage to the Sub-auxiliary shaft. The major burst was followed by a series of tremors and as many as 48 tremors were recorded by the seismic network. The computed foci of the first two major tremors lie in the hanging wall of the reef while the foci of others were located deep in the footwall of the reef and some in close proximity to the Sub-auxiliary shaft.

REFERENCES

1. Pryor, T. 1923-24. The underground geology of the Kolar goldfields, *Trans. Inst. Min. Metall., London*, vol. 33, pp. 95-115.
2. Taylor, J.T.M. 1960-61. Mining practise in the Kolar goldfields, India, *Trans. Inst. Min. Metall., London*, 70, pp. 575-604.
3. M/s. John Taylor and Sons Limited. Report of the special committee on the occurrence of rockbursts in mines of the Kolar goldfields, Mysore state, India, 1955.
4. Krishnamurthy, R. and P.D. Gupta. 1983. Rock mechanics studies on the problem of ground control and rockbursts in the Kolar goldfields. *Proc. International Sym. on Rockbursts: Prediction and Control.* IMM (Lond.), pp. 67-80.
5. Issacson, E. De. St. Q. 1957. A statistical analysis of rockbursts in the Kolar goldfields. *Bull. KGF Min. Metall. Soc.,* vol. 16, pp. 85-103.
6. Krishnamurthy, R. 1972. A review of rockburst research in the Kolar goldfields. *Proc. Sym. on Rock Mechanics* Publ. Min. Metall. Div. of Inst. Engrs. (India).
7. Bhattacharyya, A.K. 1962. Investigations into the elastic and strength properties of hornblende schists from the Kolar goldfields. M.Sc. Dissertation, Univ. of Durham.
8. Miller, E. 1965. A study of rock pressure and movements about a nearly vertical reef at great depths with particular reference to rockbursts and design of mine openings. Ph.D. Thesis, Univ. of Durham.
9. Gowd, T.N., M.V.M.S. Rao, R. Krishnamurthy, F. Rummel and H.J. Alheld. 1981. *In situ* stress measurements by hydraulic fracturing techniques in the underground mines of Kolar goldfields, India. *Proc. Indo-German Workshop. Hyderabad (India). Oct. 1981.*
10. Miller, E. Notes on rock mechanics research in the KGF, *KGF Min. Metall. Soc. Bull.,* vol. 95, pp. 23-83.
11. Sibson, J.N.S. 1967. Investigations into design and stability of workings in a deep nearly vertical reef subject to rockbursts. Ph.D. Thesis. Dept. Min. Engg., Univ. of Newcastle-Upon-Tyne.
12. Krishnamurthy, R. 1966. Strata displacement around a vertical shaft due to the mining of a protective pillar in an intersecting high inclined reef prone to rockbursts. M.Sc. Dissertation, Univ. of Newcastle-Upon-Tyne.
13. Krishnamurthy, R. and K.S. Nagarajan. 1976. Strata control measurements in stopes. *Proc. of Golden Jub. Sym. B.H.U. (India).*
14. Salamon, M.G.D. 1965. Elastic analysis of displacement and stresses induced by the mining of seam or reef deposits. Parts 1-4, *Bull. of S. Afr. Inst. of Min. Metall.,* April 1963-June 1965.
15. Murty, G.S. and R. Krishnamurthy. 1980. Proc. of Symposium on Recent Trends in Gold Mining Practice, Bharat Goldmines Limited (India) Dec. 1980.
16. Subbaramu, K.R., B.S.S. Rao, R. Krishnamurthy and C. Srinivasan. 1985. Seismic investigation of rockbursts in the Kolar goldfields. 4th Conference on Acoustic Emmision/Micro-seismic Activity in Geological Structures and Materials. Pennsylvania State University, PA.

5

VELOCITY MODEL OF CHAMPION REEF MINE, KOLAR, INDIA BELOW THE 98TH LEVEL AND ITS IMPLICATIONS FOR FUTURE PLANNING

G. Jayachandran Nair,[1] G.S. Murty,[1] Shrikant B. Shringarputale[2] and R. Krishnamurthy[2]

[1]Bhabha Atomic Research Centre, Trombay, Bombay, India
[2]Bharat Gold Mines Ltd., Kolar, Karnataka, India

INTRODUCTION

In deep metalliferous mines as well as coal mines, strata failure during mining necessitates seismic monitoring of the mining region for the safety of mining personnel. One method of monitoring strata stability is to install a network of high frequency seismic sensors near the working area to locate micro-seismic events and to use this data for planning safe mining in the region. Some instances of forewarning and withdrawal of working personnel before a rockburst using seismic precursors have been reported by Langstaff (1976) and Brady (1974). Though the physical mechanism causing the onset of fracture of rocks in deep mines is not well understood, seismic monitoring of mining regions has provided valuable symptoms of failure.

To monitor rockburst activity near the working regions of Kolar Goldfields, an array of ten sensors was installed in the Glen ore shoot region between the 98th and 103rd level. Event data from the network is obtained by a digital event recorder (Nair, 1988) installed in the laboratory near Champion Reef mine. Hard copies of events are obtained on a mingograph recorder. This paper presents an analysis of seismic records of calibration blasts conducted for evaluating the velocity model of the region and a discussion of the velocity model of

72

Glen ore shoot region obtained from such data. The model is then used to locate some events recorded during the period of experiment. The paper also addresses the issues of improving the performance and utility of the system for similar applications in future.

GEOLOGY OF KOLAR GOLDFIELDS, KARNATAKA, INDIA, AND MICRO-SEISMIC NETWORK

Kolar Goldfields in Karnataka, India is one of the oldest and deepest gold mines in Asia. The ore-bearing reef at Kolar strikes nearly N-S and dips nearly vertical as shown in Fig. 1. The mining region is almost parallel to the strike of the ore-bearing reef. The three main regions where mining activity is in progress are: (1) Mysore mine in the south where shallow mining is in progress, (2) Champion reef mine where mining has reached to depths below 3 km, and (3) Nundydroog mine in the north where the depth of mining reaches 2 km. Over the past few years it has been observed that the incidence of a large number of rockbursts is frequent near working faces in deep mines of Champion reef region. The major rockburst activity in Champion reef area is in and around the working region in Glen ore shoot. The average number of detectable events occurring per year in Champion reef area is more than 1,000. The locational accuracy of these hypocentres is better than ± 50 m. These rockbursts in Champion reef region are concentrated around working faces near Heathcote shaft and Osborne shaft of Glen ore shoot. Most of these events are induced by mining

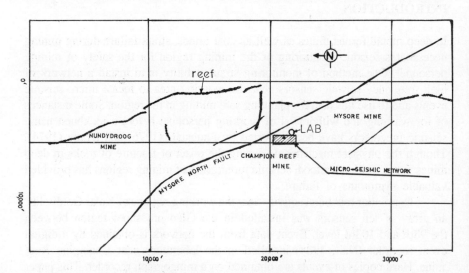

Fig. 1. Map of mining area of Kolar Goldfields showing seismogenic features, the geophone network and the micro-seismic network.

activity though some could be associated with, or be triggered by movements along Mysore North fault which cuts across the ore body near Champion reef region. In Nundydroog area, rockbursts occur at levels below 20 and are spread throughout the mining regions.

The stoping area close to Heathcote shaft, where many rockbursts occur, was chosen for the installation of a close network of geophones. In the present method of stoping, approach to the stoping region of the ore body is through footwall drives and crosscuts. Crosscuts between reef and footwall drives serve as the lifeline for working miners. Hence, the locations of geophones were so chosen as to monitor the activity behind and ahead of these stoping regions (Krishnamurthy et al., 1986). The network consists of 10 Hz geophones to record micro-seismic activity in the band of 5 Hz to 1 kHz. The plan of the micro-seismic network is shown in Fig. 2 and the location co-ordinates of the geophones are given in Table 1. Seismic signals are transmitted from the 98th level to the recording laboratory and given to the event processor unit. The unit filters the signal in the 100 Hz to 1 kHz band and digitises it by a 12-bit analogue to a digital converter. The digital data is stored in a semi-conductor memory. A description of the unit is presented by Nair (1988). For each event, the unit produces a hard copy of 0.5 sec duration starting from 0.2 sec noise prior to the onset of the signal. The paper recorder is played at 0.5 mm/msec or 1 mm/msec. During quiet periods the background noise is $\approx$ 1 mu at 100 Hz while during drilling operation it increases to 5 to 10 mu in the frequency band of 100 to 500 Hz. Events with amplitudes greater than 2 mm to 5 mm at 100 Hz are usually picked up by all sensors during quiescent periods. Events of larger magnitude from neighbouring regions are also picked up by the network.

Table 1. Co-ordinates of geophones

Location level	X kft	Y kft	Z kft
98	2191.092	0489.941	0920.365
99	2188.290	0496.812	0927.885
100	2203.615	0495.542	0935.554
101	2187.220	0492.179	0942.892
102	2204.650	0495.860	0950.125
103	2181.118	0503.447	0957.926
93	2181.120	0464.900	0882.911
100	2077.500	0449.000	0935.389
98	2180.700	0477.000	0920.400
102	2139.250	0487.950	0950.400

X, Y and Z are the x, y and z co-ordinates of the geophones.

74

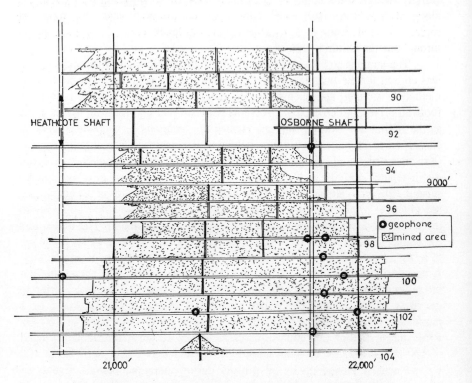

Fig. 2. Plan of the micro-seismic network in the Glen ore shoot region.

It is obvious from Table 1 and Fig. 2 that the present array is nearly planar in the X-Z plane and due to the geometry of mining it is difficult to install geophones in the Y direction, i.e., perpendicular to the footwall to improve source location accuracy. Also, for on-line processing of data the onsets should be impulsive and the precision of onset estimation should be accurate enough to give the hypocentral location with an error radius of 3 to 4 m. An alternate way of analysing data is to record and replay them continuously; for such purpose one requires data recorders with large bandwidth and large storage memories. As the noise portion of the records is large compared with the signal portion, editing input data is necessary before processing them in real time. Such an approach to signal processing of voluminous seismic data is currently adapted in on-line analysis of seismic arrays (Key, 1976). With this aim in view, an event processor unit which will detect and buffer events for further processing has been developed. The unit also serves as an interface to the ADC unit of the computer for processing event portions of the records and extracting relevant parameters in real time.

VELOCITY CALIBRATION OF MINING REGION

Analysis of primary hard copy data of blasts in underground working regions in Kolar goldfields by the commonly used least-square programmes (Crosson, 1976) showed that to determine each location accurately, a heterogeneous and anisotropic velocity model of the region should be used. For this purpose, a series of blasts at known locations near the working areas was carried out in order to calibrate each region. The locations of the blasts and the amount of explosives were decided according to mine safety requirements. Planning of the blasts was carried out with the expertise of the BGML staff. Survey locations of the blasts were provided by the Survey Department of BGML. Origin time was computed using known velocity and distance of the blast to the nearest geophone as this was less than 10 m. The onset time of the P-wave at each

Table 2a. Velocity models obtained from iteration

Trial no	Velocity (kft/s)			Estimated Hypocentre (kft)			Errors		
	Vx	Vy	Vz	X	Y	Z	DR (ft)	Dt (msc)	Tor (msc)
Shot location 1				(21930)	(4906)	(9201)			
1	19.0	11.4	19.8	21924	4904	9199	6.3	0.14	+0.14
Shot location 2				(21855)	(4948)	(9275)			
1	19.0	10.0	20.0	21867	4938	9269	28.0	0.6	+1.0
2	20.2	10.0	18.0	21854	4936	9274	11.7	0.6	+1.7
Shot location 3				(22024)	(4940)	(9351)			
1	19.5	10.0	18.0	22022	4938	9344	7.5	0.45	1.9
Shot location 4				(21891)	(4923)	(9425)			
1	19.0	10.0	18.9	21895	4943	9427	20.5	0.47	−0.14
Shot location 5				(22035)	(4945)	(9497)			
1	19.0	10.0	19.0	22033	4942	9499	4.2	0.23	+0.24

Vx, Vy, Vz are the compressional wave velocities in x, y and z directions. X, Y, Z are the x, y and z co-ordinates of the hypocentre. DR and Dt are error residuals in ft and milliseconds. Tor is the origin time in msec. Values in parentheses are the actual co-ordinates.

Table 2b. Estimated hypocentres using velocity model in Table 2a

Trial no	Velocity (kft/s)			Estimated Hypocentre (kft)			Errors		
	Vx	Vy	Vz	X	Y	Z	DR (ft)	Dt (msc)	Tor (msc)
Shot location 6				(21855)	(4948)	(9276)			
1	20.2	10.0	18.8	21856	4933	9276	15.0	0.6	+1.76
Shot location 7				(22099)	(4899)	(9355)			
1	19.5	10.0	18.0	22099	4897	9353	2.8	0.3	+5.7

Vx, Vy, Vz are the compressional wave velocities in x, y and z directions. X, Y, Z are the x, y and z co-ordinates of the hypocentre. DR and Dt are error residuals in ft and milliseconds. Tor is the origin time in msec. Values in parentheses are the actual co-ordinates.

sensor was read from the triggered hard copy playouts to an accuracy of 0.5 msec. An iterative programme which best fits the hypocentre and origin time along with an anisotropic model was run for the five blasts at levels 98, 99, 100, 101 and 102. Sample records of the triggered blasts for levels 99, 100, 101 and 102 are given in Figs. 3 and 4. The best fit velocities for five levels, in shot location levels 1 to 5 are given in Table 2a along with details of the shot locations. These velocities are further used to locate known blasts from neighbouring regions. Computed and actual locations for two blasts among the twenty blasts analysed are given in Table 2b for shot locations 6 and 7. It is seen that the relocated location of these events with calibrated velocities agree with actual values within 4.5 m.

FUTURE PLANNING FOR EXPANSION AND UTILIZATION OF THE SYSTEM

For forecasting impending rockbursts, acoustic emissions occurring before actual failure should be located with a fair amount of accuracy. In this light, observed heterogeneity and possibility of time-varying heterogeneity in the medium should be monitored in order to study the stability of the mining region. Hence, expansion of the existing network as well as monitoring of other associated properties such as attenuation become necessary.

Modelling of fracture as time-dependent topology was shown to produce changes in compressional and shear velocities prior to failure in rocks (Aggarwal et al., 1975). It is also seen that in heterogeneous medium, changes in the topology defects also change the Q (quality) factor of the medium. The void volume

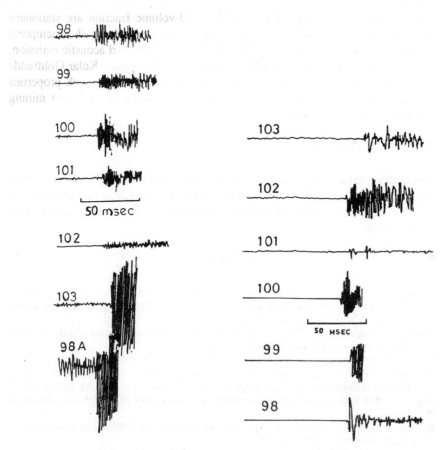

Fig. 3. Triggered record of calibration blasts at levels 98 and 99.

Fig. 4. The triggered record of calibration blasts at levels 101 and 102.

fraction reaches maximum and then decreases before actual failure in a heterogeneous medium. A study of the effect of compressional speed and ratio of Dt^*/t^* (where $t^* = tp/Q$; tp-compressional wave travel time and Q quality factor; and Dt^*-incremental value of t^* with void volume) with void volume showed that near void volume values close to failure the ratio Dt^*/t^* could be a better index of failure than the commonly used t_s/t_p criteria, where t_s is the shear wave travel time (Murty and Nair, 1981).

In order to study experimentally the relative utility of these theoretical ideas it is necessary to generate controlled acoustical pulses at known locations periodically and to monitor the variation of compressional speed as well as attenuation properties of these waves traversing a potentially unstable region. If the medium properties remain stationary during the experiment, then we can conclude that the state of stress and the void-volume fraction are stationary and hence the region is stable. If, on the contrary, these parameters show temporal variations, then observing them along with accurately located acoustic emission, one could possibly predict the future behaviour of the region. Kolar Goldfields is the most ideally suited region for such controlled study of rock properties *in situ*, which could contribute in the long run to studies in reducing mining hazards in similar regions.

ACKNOWLEDGEMENTS

The authors wish to thank all the technical and scientific staff in Gauribidanur Seismic Array Station, Karnataka and Bharat Gold Mines Ltd. who were involved in the collaborative project between BARC and BGML for their contribution in this work.

REFERENCES

Aggarwal, Y.P., L.R. Sykes, D.W. Simpson and P.G. Richards. 1975. Spatial temporal variation in t_s/t_p and in P-wave residual at Blue Mountain Lake, New York: Application to earthquake prediction *J. Geophys. Res.*, vol. 80, pp. 718-732.

Blake, W., F. Leighton and W.I. Duvall. 1974. Microseismic techniques for monitoring the behaviour of rock structures. *Bu Mines Bull.* No. 665, 65 pp.

Brady, B.T. 1974. Seismic precursors before rock failures in mines. *Nature*, vol. 252, p. 5484.

Crosson, R.S. 1976. Crustal structure modeling of earthquake data. 1. Simultaneous least squares estimation of hypocentre and velocity parameters. *J. Geophys. Res.*, vol. 81, pp. 3036-3046.

Key, F.A. 1976. Digital processing of seismometer array data, SSD, AG 197, UKAE, Blacknest, Brimpton, U.K.

Krishnamurthy, R., K.R. Subbaramu, B.S.S. Rao and A.G. Kulkarni. 1986. Micro-seismic network at Kolar goldfields, *BARC*, no. I-881.

Langstaff, J. 1976. Helca-seismic detection system. *Proc. 17th Rock Mechanics Symposium, Snowbird, Utah.*

Murty, G.S. and G.J. Nair. 1981. Fracture as a time-dependent topology of inclusions. *International Symp. on Rock Mechanics*, NGRI, Hyderabad.

Nair, G.J. 1988. PDP 11/34-based rockburst monitoring system at Kolar goldfields. *Ibid.*

6

REVIEW OF WORKING OF SEISMIC AND MICRO-SEISMIC NETWORK INSTALLED AT KOLAR GOLDFIELDS

K.R. Subbaramu[1] and R. Krishnamurthy[2]

[1]Bhabha Atomic Research Centre, Gauribidanur
[2]Bharat Gold Mines Ltd., KGF

INTRODUCTION

Investigations concerning the nature and cause of rockburst are of great significance to the safety and economics of mining operations in deep mines such as in Kolar Goldfields.

Instruments used at Kolar Goldfields prior to 1978 consisted mainly of extensometers, stressmeters, convergence meters, packload cells etc., which had certain limitations, the main one being that they could only detect changes in stress and strain at their location. As a major stoping sequence may extend over a large area, it is impracticable to instrument the entire area.

Rockburst, a consequence of changes in stress pattern caused by extensive mining, releases seismic energy. This release of energy can be monitored using seismic techniques which provide data for controlling mining activity.

Monitoring of seismic activity in areas prone to rockbursts has become a standard practice in most deep mines around the world. Trial seismic monitoring had been undertaken by KGF authorities prior to 1978 using instruments provided by the University of Newcastle-upon-Tyne, U.K. [1]. The mode of recording had some limitations. In 1975, on the initiative of Bharat Gold Mine authorities and financial assistance from the Department of Science and Technology, Government of India, a joint collaborative project was started with Bhabha Atomic

Research Centre to set up a seismic network [2] to cover the entire mining area, the main objective of which was to locate sources as accurately as possible and to establish a velocity model for the entire area. The setup was established in 1978. Prior to this, two preliminary recordings had been done by BARC personnel in 1969 and 1971. The present system in operation used this experience to improve the instrumentation setup.

In phase II, a close-in micro-seismic network [3] was established in a working mine in 1983. The object of this exercise was to record rock movements which might eventually result in a major rockburst.

The seismic and micro-seismic networks for monitoring all types of rockbursts have been operating continuously since then were set up. The basic plan of such instrumentation was that it be simple and easy to maintain utilising trained personnel from BARC laboratories. Fabrication of instruments was also planned in such a way that except for magnetic tape drives and paper recorders, all electronic units could utilise components available indigenously.

REVIEW OF EXPERIENCES

Initially it was proposed that the eight sensors in Phase I be set up in an 'L' shape as this best suited the mining area. Later, on the basis of theoretical studies by Arora and Basu [4], a hexagonal array covering the entire mining region was proposed and actual sensor locations were determined after extensive reconnaissance of the area and *in situ* noise measurements.

Before the final installation, trial recordings of four seismic channels were initiated at the locations on a single channel audio tape recorder using PPM recording technique. Data thus obtained gave useful hints for incorporating improvements in the final installation.

Many options were considered for telemetring the signals. Finally, it was planned to have part of the array telemetred by cable and the balance using wireless. Indigenous firms were asked to manufacture cables to the required specification. For wireless telemetry, an indigenous trans-receiver was utilised. So that more than one seismic signal could be transmitted on a single carrier frequency, we used the pulse position modulation (PPM) technique [5].

Various PCBs have been designed on the basis of standard electronic circuits with certain modifications incorporated to suit site conditions. All PCBs have been standardised to one size so that a single type of modular unit can house them. The PCBs are the plug-in type with test points brought out. Components for the various circuits were so chosen as to withstand wide-ranging environmental conditions. Integrated circuit chips were soldered directly on to the PCB to avoid contact problems. All the edge pins were gold plated for good contact.

Initially, for about two years, the entire recording was done on a single shift basis. From operational experience of the maintenance staff and dialogue with the data users, many changes were incorporated so that the system could be operated continuously without much trouble. Because of added flexibility in the design of preamplifiers, frequency modulator, demodulators, filters, etc., frequency response, amplifier gain and carrier frequency could easily be altered in the field depending on the requirement of the data user for acquiring good quality data. To begin with, five direct cable telemetred and three wireless telemetred channels were installed. The signal band width was restricted to 5 to 50 Hz for cable telemetred channels and 5 to 30 Hz for wireless telemetred channels. Later, the wireless channel band width was also changed from 5 to 45 Hz. In spite of these changes, raise times recorded from the signals did not improve markedly, resulting in some error in location parameters. Moreover, the data transmitted by wireless was not up to the mark. This was mainly due to limitations of the trans-receiver set. Ultimately, after about four years of operation, the wireless transmission was abandoned and replaced by cable. PPM signals were directly used to couple long cable lengths using line drivers. Due to pickup problems, the PPM technique was discontinued and FM signals directly driving the cable used. Further, the number of channels was increased from 8 to 14. Despite these changes, locational accuracy was still not good enough. It was decided to increase the frequency response from 5-50 Hz to 5-170 Hz. This improved the location considerably and the locational accuracy now obtainable is of the order of ± 25 metres. After the first two years of operations, a continuous three-shift operation of recording was begun and still continues. Over the past nine years breakdowns have been minimum. Only on one or two occasions has stoppage of recording occurred. This was due to some lightning damage and human vandalism. Some problems with the paper recorder and tape deck were also experienced but resolved without difficulty.

Phase II instruments, set up in 1983, differed from Phase I in mode of recording. This was mainly because transducers were located closer to a working mine from which we expected more events, both big and small, with a higher frequency content. So the technique of registering them was changed. Instead of recording on a magnetic tape, events above a certain threshold were counted and their time plot used to monitor mining areas. Signal transmission from underground levels used the same type of cable as used in Phase I. Even the field electronics was similar to the one used in Phase I. Only the frequency response of the amplifier and carrier frequency for frequency modulation differ. Sensors have better HF response, a basic requirement because of close proximity to the source. Initially an accelerometer was installed. But initial trials posed some problems. So accelerometers were replaced by HF geophones which could record frequencies up to 800 HZ. Before installing the sensors at deeper levels, a noise survey was conducted. Based on actual noise data, sensor locations were fixed. Also, from preliminary recordings the frequency response for the setup was fixed at 200 Hz-800 Hz.

This setup has been in continuous operation since its installation. There was one problem of a frequency modulation carrier in the field drifting because of high temperatures underground. It was corrected using high stability capacitors with a good temperature coefficient. A tricky problem faced in Phase II recording was obtaining reliable data during drilling operations. Methods of filtering in various bands were tried for elimination of drilling noise but eventually a compromise was struck. Since drilling operations are known before hand the best possible band pass filter setting was obtained experimentally. Nevertheless, obtaining reliable data during drilling remains a problem to be tackled. From published literature on the subject, apparently similar problems are faced in other mines also [7].

Over the years, the performance of various instruments has been constantly assessed. A calibration test was performed on one typical setup utilising the shake-table facility available at one of the laboratories in Bangalore. The results obtained accorded with design parameters [6]. A few calibration shot experiments from known locations were also conducted. Shot instant was recorded separately.

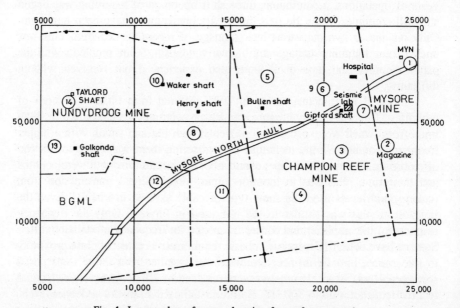

Fig. 1. Layout plan of sensor location for surface network.

From this, the arrival time at various sensors was calculated from which a velocity model was derived.

Some results obtained with the surface network are given in Tables 1, 2 and 3. Table 1 gives an overview of the number of rockbursts recorded from 1980 to 1984. Table 2 shows the values of velocity derived from the data in X, Y and Z directions for the different mining areas. Table 3 projects the number of rockbursts having different energies recorded during 1983 and 1984. Table 4 gives the chronological order of project development from 1975 to 1987. Figures 1 and 2 show the plan of surface seismic network and micro-seismic network. Typical rockburst wave forms are shown in Fig. 3. Figure 4 gives a picture of foci of rockbursts plotted as a function X and Z co-ordinates of the mines. These were obtained from the surface network recording. Figure 5 is a typical micro-seismic record plotted for a 24-hour period.

OSBORNE SHAFT

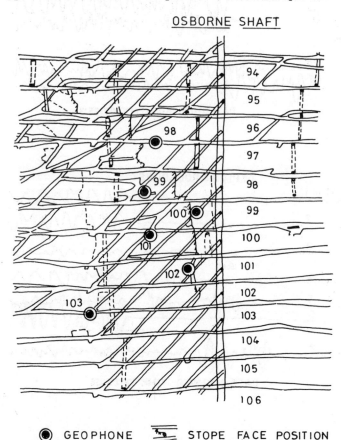

⬤ GEOPHONE   STOPE FACE POSITION

Fig. 2. Layout plan of sensor location for underground network.

84

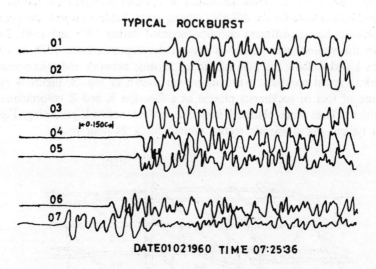

TYPICAL ROCKBURST

01
02
03
04
05

06
07

DATE01021960 TIME 07:2536

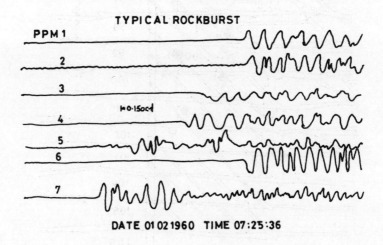

TYPICAL ROCKBURST

PPM1
2
3
4
5
6

7

DATE 01 02 1960 TIME 07:25:36

Fig. 3. Typical rockburst waveforms.

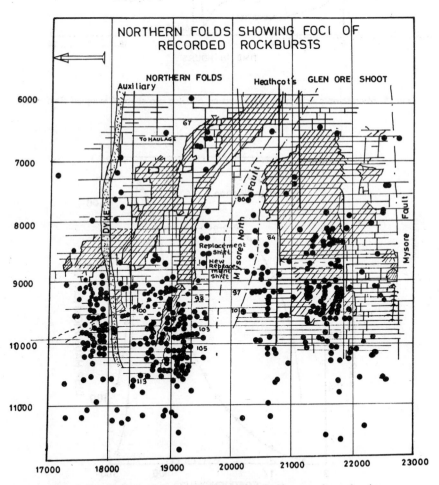

Fig. 4. Longitudinal section of Northern folds & Glen ore shoot showing
foci of recorded rockbursts.

86

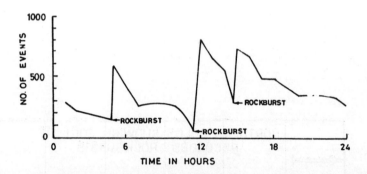

Fig. 5. Microseismic activity in south section of Glen ore shoot, Champion Reef Mine.

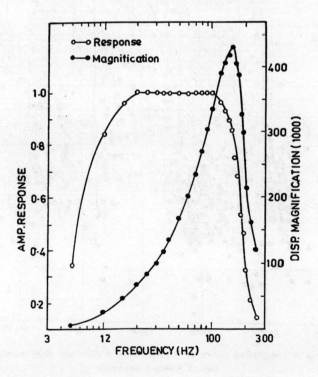

Fig. 6. Response characteristics of the electronic system.

Figure 6 shows the response characteristics of the electronic system and displacement magnification curve obtained by shake-table calibration.

Table 1. Number of rockbursts recorded during the period 1st January 1980 to 31st December 1984 and number reported from various mines

Year	No. of rockbursts recorded by seismic network	No. of rockbursts reported from mines
1980	688	10
1981	908	10
1982	1,453	20
1983	848	16
1984	1,354	26

Table 2. Velocity model derived from seismic data

Area	Velocity in km/sec		
	V_x	V_y	V_z
Champion Reef Mine:			
Northern Folds (East limb)	6.2	5.3	6.6
Northern Folds (North of dyke)	6.1	5.4	6.2
Glen Ore Shoot (North Wing)	6.2	5.4	6.6
Glen Ore Shoot (South Wing)	6.2	5.4	6.3
Nundydroog Mine	6.6	6.45	6.5

Table 3. Energy liberated by rockbursts recorded in KGF by seismic network

Year	Energy in foot pounds	10^4	10^5	10^6	10^7	10^8	10^9	And above
1983	Number of events	60	314	281	147	42	4	
	Instances of damage	—	—	5	7	4	4	(All these were in inaccessible areas)
1984	Number of events	75	341	464	464	68	10	
	Instances of damage	—	—	6	13	5	10	(8 were in inaccessible areas)

Table 4. Project development-chronological sequence

October 1975	Agreement on project proposal for Phase I.
October 1975 to June 1976	Drawing up to specifications. Noise measurements.
July 1976 to September 1976	Fabrication of prototype units. Trials with wireless transmission.

(Contd.)

Table 4 *(Contd.)*

October 1976 to December 1976	Testing of prototype field units in the field. Improvements in pulse position modulation techniques.
January 1977 to March 1977	Three-element geophone array field tested using cable and wireless telemetry for a period of 4 weeks. Tested head driver unit for TD 10 analogue tape deck. Work on replay electronics initiated. Fabrication of PCBs for field electronics.
April 1977 to June 1977	Wireless telemetry using PPM technique field tested over a distance of 8 km. PCBs for different recording systems fabricated.
July 1977 to September 1977	Wireless telemetry field tested at KGF for a period of 40 days. During this period some rockbursts and detonation of chemical charges were recorded. Work on filter and shaper unit initiated.
October 1977 to December 1977	Fabrication and testing of FM head driver, replay amplifier, frequency modulation demodulator, time-code generator and reader completed. Filter and shaper unit developed. Modular units to house various PCBs designed.
January 1978 to March 1978	Modifications carried out to improve performance of PPM telemetry. Work of filter and shaper completed. Laying of cable for establishing partial network at KGF taken up. Theoretical study of triangular network using simulated data started. Computer programmes written and tested.
April 1978 to June 1978	Based on experience gained by multichannel trial recordings on audio tapes using PPM techniques, surface sensors increased from envisaged 7 Nos to 11 Nos. Work on all electronic units completed. Final tests and calibration undertaken.
July 1978 to September 1978	Fabricated and tested units moved to the site in August 1978. Installation of eight channels (6 direct and 2 PPM) completed by 8th September 1978. System worked during a single shift.
October 1978 to December 1978	Rockbursts recorded at KGF replayed and analysed at Trombay using BESM computer and some analysed using a programmable calculator acquired for the project. It was established that data of BESM-6 and programmable calculator agreed.
January 1979 to March 1979	Assessed working of instruments after installation. All differences in gain and frequency responses made identical; receiver installed for time correction. Incorporated remote command sine-wave calibration facility. Installed tape replay system. Results of analysis of rockbursts indicated anisotropic velocity distribution. This prompted installation of at least 3 extra geophones at deeper levels. Work on this was initiated by BGML.
April 1979 to December 1981	Constant reviewing of working of instruments. Remedial actions for faults observed taken. Velocity model obtained. Location for a number of rockbursts calculated using derived velocity model for the area.
October 1981	Beginning of Phase II Work.
October 1981 to April 1982	Planning for instrumentation discussed. Work on field amplifier, power supply, delay processor started. Preliminary noise survey conducted. Tests on suitability of cable for telemetry conducted. Contacts established with various firms to assess suitability of a micro-processor system.

(Contd.)

May 1982 to May 1983	3-shift operation continued. Rockburst statistics obtained. Some migration study of rockburst done. Computation of stress concentration using rockburst-duration method initiated. In-house maintenance carried out regularly.
May 1983 to December 1983	Laying of cables for Phase II instruments completed in Gifford shaft. Fabrication of instruments for Phase II continued. HF signals from close-in-network located around Osborne shaft south stope continuously recorded.
January 1984	Event counter installed. Problem in signals from level 102 cleared. Calibrated monitor recorder and mingograph recorder. Replay noise reduced. Laying of cable to 70th level Gifford shaft, Champion reef completed.
February 1984 to March 1984	Experiments conducted to find frequency rate of drilling noise signal.
April 1984	HF signals recorded by micro-seismic network used for correlating mining and other activities in the region. Attempts made to replace HF geophones with accelerometers.
May 1984 to June 1984	Micro-seismic recording continued. Response of filter and shaper circuit modified to accommodate accelerometer signals.
September 1984	Recording of micro-seismic signals continued. All geophones replaced by accelerometers. Uptron Micro-processor Software being developed.
October 1984	HF signals from close-in network of accelerometers continued. Modification done in order to improve quality of signal recording. Interface required for micro-processor and software development work continued. Fabrication of delay processor in progress.
November 1984	Close-in recording continued. Modifications effected in electronic circuit to improve quality of accelerometers and HF geophone signals. Experiments conducted to check performance of accelerometers and HF geophones at different electronic gain and filter settings. Delay processor fabrication continued.
December 1984 to January 1985	Close-in network recording continued. Blasting signals were recorded and replayed. Using this data, velocity model derived. Micro-seismic emission rates plotted continuously. Software development for micro-processor continued.
March 1985	Installation of micro-seismic network completed. Fault developed at level 103 rectified. Data collection continued.
March 1985 to April 1985	Data recording continued. PDP 11/34 mini-computer installed.
April 1986 to August 1987	PDP 11/34 used for analysing data from micro-seismic network. S-850 micro-processor installed and used for online analysis of data.

CONCLUDING REMARKS

Seismic monitoring of rockbursts occurring in and around mining regions has proved very helpful in assessing the stability of stoping areas and controlling mining operations, depending upon seismic activity recorded from the area. It has been possible from this technique to demarcate high-stress zones.

Micro-seismic monitoring has indicated an increase in micro-seismic activity followed by a lull prior to a local burst (Fig. 6). Data obtained by micro-seismic recording along with strain measurement are used for assessing day-to-day stope stability to issue forewarning of any impending danger.

It may be added that the success of this project can largely be attributed to close contact and constant discussion between the users and planners of the instrumentation system. At every stage of the project, there has been continuing dialogue to improve the instruments and update them. Further, the success of the project may also be attributed to the rigorous training given to BGML personnel during fabrication, testing and installation.

ACKNOWLEDGEMENTS

The authors are indebted to the authorities of BGML and BARC for extending all possible help to implement the project successfully. Acknowledgement is also made to all the personnel of BGML and BARC who participated and gave wholehearted support in executing the project.

REFERENCES

1. Bhattacharyya, A.K. 1967. Application of seismic technique to problems in rock mechanics. Ph.D Thesis. University of Newcastle-upon-Tyne.
2. BARC-BGML Scientists. 1980. Seismic investigation of rockbursts in Kolar goldfields. KGF centenary celebrations. Seminar Proceedings: Recent Trends in Gold-Mining Practise. December.
3. Krishnamurthy, R. (BGML), B.S.S. Rao, A.G. Kulkarni, K.R. Subbaramu (BARC). 1984. A network to monitor micro-seismic activity. Symposium on Role of R & D in Indian Mining Industry Conducted by Institute of Engineers, Bangalore Chapter.
4. Arora, S.K. and T.K. Basu (BARC). 1978. Relative performance of different triangular networks in locating regional seismic sources, *Proceedings of Indian Academy Science,* vol. 87A, No. 11, November.
5. Syam Sunder Rao Bora (BARC). 1980. Communication systems: An application to seismology. KGF centenary celebrations. Seminar Proceedings: Recent Trends in Gold-Mining Practice, December.
6. Arora, S.K., B.S.S. Rao, A.G. Kulkarni, K.R. Subbaramu (BARC) and R. Krishnamurthy (BGML) 1983. Calibration of the seismic system for detection of rockbursts at Kolar Goldfields. *BARC,* No. I-760.
7. Brink, A.V.Z. and D.M.O. Connor. 1983. Research on the prediction of rockbursts at western deep levels, *Journal South African Institute of Mining and Metallurgy,* January 1983.
8. Rao, B.S.S., K.R. Subbaramu (BARC), R. Krishnamurthy and C. Srinivasan (BGML). 1985. *Proceedings of Fourth Conference on Acoustic Emission/Micro-seismic Activity in Geological Materials* held at Pennsylvania State University, June.

7

PDP11/34-BASED ROCKBURST MONITORING SYSTEM FOR KOLAR GOLDFIELDS

G. Jayachandran Nair and V.S. Kamath

Bhabha Atomic Research Centre, Bombay-85

INTRODUCTION

In deep metalliferous mines and coal mines the hazards due to strata failure during mining warrant seismic monitoring for the safety of mining personnel. This is usually accomplished by deploying a seismic surveillance system with a network of high-frequency sensors near the working area to locate micro-seismic events and to forewarn instability of the mining region. Such a system usually requires real-time data acquisition and processing capability. Description of the high-frequency sensor network and associated electronics for micro-seismic data acquisition is available in literature (Langstaff, 1974; Blake *et al.*, 1974). Instances of increase and decrease in micro-seismic activity prior to large rockbursts have been observed in zinc mines and silver mines. In some instances, withdrawal of working personnel from the mining area prior to the rockburst was possible using seismic precursors (Langstaff, 1976; Brady *et al.*, 1974). Although the physical mechanism causing the onset of fracture of rocks in deep mines is not well understood, seismic monitoring of a mining region has provided valuable symptoms of impending failure.

Commercially available medium-size computers with data interfaces such as ADCs are capable of directly handling data from two or three networks with each network having 10 to 15 geophones (Leighton, 1983). This paper presents the technical and functional details of an event data interface unit for enhancing the data acquisition capability of a monitoring system to 15 to 20 networks, using a common data cable. The digital event data interface unit interrupts the processor when an event occurs in a network and transfers the buffered data of the event to the computer. Such a unit is currently deployed in the Glen ore shoot region of the Kolar Goldfields, Karnataka, India for gathering micro-seismic event data for on-line estimation of event parameters by a PDP 11/34 computer. This paper also discusses some events recorded by the unit.

KOLAR GOLDFIELDS KARNATAKA, INDIA AND MICRO-SEISMIC NETWORK

Kolar Goldfields in Karnataka, India is one of the oldest and deepest gold mines in Asia. The ore-bearing reef at Kolar strikes nearly NS and the main region of mining parallels the direction of the reef strike. The mining area consists of three regions—Mysore mine, Champion Reef mine where high grade ore is mined at depths of 3 km and Nundydroog mine. The majority of rockbursts in Champion Reef mine have concentrated around the working faces near Heathcote shaft and Osborne shaft. Most micro-seismic activities are associated

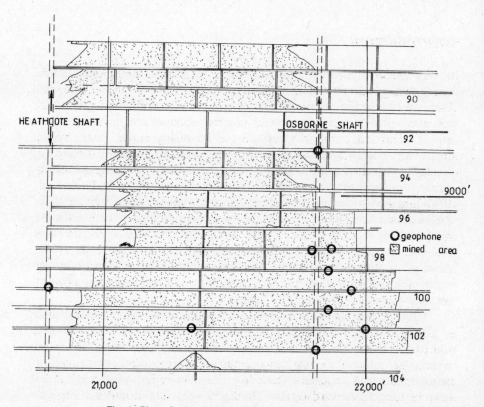

Fig. 1. Plan of micro-seismic array near the Osborne shaft.

with excavation of the ore body while some of these events could be associated with, or triggered by, the Mysore North fault.

As most of the recorded rockburst activity near Champion Reef mine has concentrated in the region of Glen ore shoot, this region was chosen for the installation of a close-in network of geophones. The plan of the micro-seismic geophone array is shown in Fig. 1. The network consists of 10 Hz geophones with a band width of 5 Hz to 1 kHz. The output of the geophone is amplified by about 40 to 60 dB depending on the site noise and is modulated on a 12 Hz carrier for transmission to the surface laboratory via a multi-core cable. In the laboratory, this signal is recovered by demodulating the carrier and then fed to the event processor unit for detection and digitisation. During quiet periods the background noise is 2 mu in the 100-500 Hz frequency band. During drilling operation, the noise increases to 5-10 mu. Near events with signal amplitudes greater than 5 mu in the frequency band 100 Hz-1 kHz are usually picked up by all sensors during quiescent periods. Signals from distant rockbursts are also picked up by the network and using the onset times or frequency rate alone, they cannot be distinguished from those occurring within or close to the network. Hence several more geophones were installed at greater distances to remove this ambiguity.

The present array is nearly planar; the geometry of the mine does not permit installation of geophones in the Y direction, i.e., perpendicular to the footwall, to improve array locational accuracy. For on-line processing of the data, the onsets of the signals should be impulsive and the precision of onset estimation should be accurate enough to compute the hypocentres with an accuracy of 3 to 4 m. However, for recording and replaying this data, an on-line processing system with large band-width data acquisition and storage capacity is required. As the noise portion of the records is large compared with the signal portion, processing of the edited event portions of the data is preferable to continuous on-line processing. With this aim in view, an event processor unit which detects and buffers the data of events for on-line processing has been developed. The unit serves as an interface to the ADC of the computer for transmitting the event portions of the records for extracting the source parameters.

EVENT PROCESSOR UNIT

The functions of an event processor unit are filtering of micro-seismic data, buffering event portions of the data and sending the data to the computer ADC on interrupt for acquisition and processing. The unit accepts data from a cluster of eight geophones and stores past data of all channels for a time duration of 500 msec in a cyclic buffer of 64 kb length. When an event is registered, this data is sent on interrupt to the computer. A hardware circuitry is provided for registering an event when coincidence occurs in a preset number of channels.

The unit consists of a pre-filter, which filters the data in the required pass band, an event detector circuit for triggering the event portion of the data, a coincidence circuit which triggers the coincidence of event in a preset number of channels, a digitiser and buffer memory for preserving the past 500 msec of data of all channels and an interrupt circuitry which transmits the data to the computer when an event is registered. The block diagram of the unit developed for this purpose is shown in Fig. 2. The sub-units of the system are pre-filter, event detector and digital data interface unit.

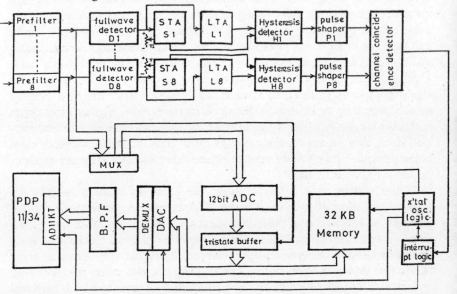

Fig. 2. Block diagram of the event data interface unit.

Pre-filter

The analogue data of the eight sensors after demodulation is fed to this unit for improving the signal-to-noise ratio. The sub-unit provides a pass band of 25 Hz to 1000 Hz using an eighth order Butterworth active filter with a sharp 120 db cutoff. The corner frequencies of the pass band are hardware adjustable. The field data are passed through this unit before further processing for removing dc and for improving the signal-to-noise ratio.

Event Detector Unit

Filtered data is fed to the full-wave detector D1 and passed through the short-term averager (STA) S1 and long-term averager (LTA) L1. The time constants of the short-term averager and the long-term averager are chosen as 20 msec and 200 msec respectively. The normalised long-term average is added with a variable threshold in T1 and compared with the short-term average in a hysteresis comparator H1 for presence of a signal above the noise hysteresis. When the signal is present, the output of H1 remains high until the signal energy falls below the background noise. When the signal energy falls below this threshold, the output is low, indicating the presence of noise only. Thus the output of the hysteresis amplifier is +10 v for no signal and becomes −10 v when a signal is present. This output is fed to monostable M1 to produce a pulse of 200 msec on the transition of the hysteresis output. Thus when an event signal arrives in each channel, one 200 msec TTL level pulse is generated per channel. The maximum distance between the sensors is about 400 m and for an average P-wave velocity of 5.0 km/sec the maximum time lag between any pair of sensors will be 80 msec. These pulses from different channels are added and the summed output reveals the pulse voltage proportional to the number of channels in the overlapping region, whenever an event is present. A summing resistor network is used to sum these pulses from different channels and this summed voltage is compared against the voltage from a selector which produces the required voltage for coincidence in the switch-selected channel number. A TTL level signal is produced at the output of the comparator whenever coincidence occurs in the pre-set number of channels. This TTL signal E1 is sent to the digital data interface unit in order to open the gate of data to the PDP 11/34 computer, as well as to produce interrupt logic pulses for accepting the data into the computer.

Digital Data Interface Unit

The digital data interface unit digitises the filtered analogue data and stores it in a cyclic buffer of 32 kilowords of 12 bit length. When an event pulse E1 is received at time TO by the unit, the unit transfers the past stored data whose time axis starts from TO-0.5 sec to the computer for further processing. The unit stores data from eight channels and has provision for providing digital as well as analogue data of events starting about 0.5 sec prior to the onset of the event. As the output data is available in cosmos tristate logic, the unit is capable of mixing data of to 8 to 16 similar units and giving the multiplexed data to a common transmission path or common input port of a computer. This means that event data from 8 to 16 regions can be monitored by a single computer if the collision probability of the events data packet is small. The block diagram of the unit shown in Fig. 2 shows that the 8-analogue channels are muxed with a slew rate of 4 kHz and given to the input of a 12-bit ADC. The logic pulses for muxing, digitising and other logic operations are derived from a 2 MHz clock pulse by division and decoding. Pulse shapers are used to shape as well as to delay pulses for the logic operations. The digital data from the ADC is buffered in a tristate buffer and interfaced to a Mos memory bank

96

16-bit 32-kiloword long. The read-write pulses are obtained from the crystal clock logic and the address of the memory is taken from a dual four-bit binary counter. For chip select a binary counter with decoder is used. The written data is read with a variable shift. The maximum shift occurs when the same location is read and then written with the new data which provides a time delay of 0.5 sec. The read data is strobed into a 12-bit buffer and converted into analogue form by a DAC with current feedback loop. The output of the DAC is followed by an op-amp for gain adjustment and dc compensation. The address of the channel is taken from the input mux and the analogue output is demuxed using a Mos tristate demultiplexer. The demuxed output is buffered by an op-amp voltage follower and given as the output of the channel. This analogue output is given to the input of the ADC of the computer through a multi-feedback band pass filter with the same characteristics as the input filter. For interrupting the PDP 11/34 AD11KT analogue to digital converter, a TTL level down-going pulse of 40 to 60 msec duration is generated when the event pulse is registered by the interface logic. This interrupt pulse occurs at a set sampling frequency for a duration of 0.5 sec. Sampling frequency and duration of flow of data are variable.

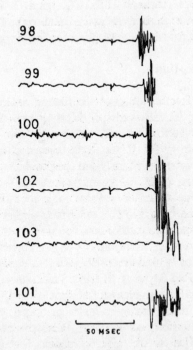

Fig. 3. Sample records of blasts in the 98th level.

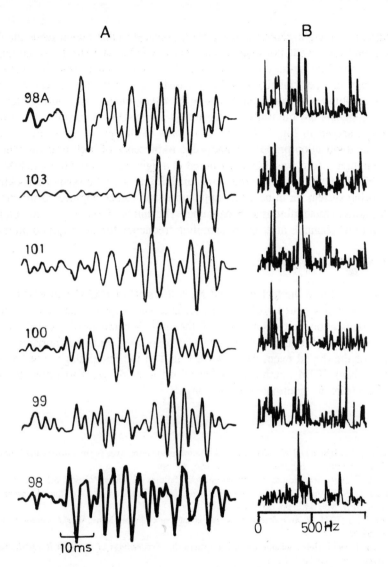

Fig. 4. Digitally filtered computer reproduction of event records in Fig. 3. Trace A corresponds to signal and trace B corresponds to fourier spectrum of trace A.

OPERATION AND SAMPLE EVENT RECORDS OBTAINED FROM THE SYSTEM

The system was put into operation from late 1985 for monitoring micro-seismic activities near the Glen ore shoot region in the underground mines of Kolar

Goldfields. An eight-geophone network established in this region feeds the data to the unit and event portions of the data are fed to a PDP 11/34 computer in real time. The system with the PDP 11/34 has been operational for over a year and its operational performance satisfactory. It is able to record events whose amplitudes are appreciably above the background noise. During blasting, all blast signals are faithfully picked up by the system. A sample of the blast record is shown in Fig. 3. Computer records of this event are shown in Fig. 4. As the digitisation produces signals due to presence of high frequency noise, the computer record was digitally filtered in a tight band from 50 Hz to 800 Hz.

The system has demonstrated its capability as a pilot model of a rockburst monitoring system for mine safety application and can be engineered for similar applications in metalliferous and coal mines. The unit has flexibility for monitoring data from 64 channels at different sampling frequency for any required duration.

ACKNOWLEDGEMENTS

We are thankful to the technical and scientific staff of BARC and BGML who were actively associated with this project and who helped in achieving success in the fabrication and installation of the unit. We are also grateful for the contributions of Mr. D.S.S. Rao who was associated with the fabrication and testing of the digital event data interface unit in the initial stage. We likewise wish to thank BGML authorities who helped in many ways to install the unit and to ensure its continuous operation.

REFERENCES

Blake, W., F. Leighton and W.I. Duvall. 1974. Microseismic techniques for monitoring the behaviour of rock structures, *Bu Mines Bull.*, No. 665, 65 pp.

Brady, B.T. 1974. Seismic prescursors before rock failures in Mines, *Nature*, vol. 252, p. 5484.

Brady, B.T. 1977. Anomalous seismicity prior to rockbursts: Implications for earthquake prediction, *PAGEOPH*, vol. 115, pp. 357-374.

Langstaff, J.T. 1974. Seismic detection system at the Lucky Friday Mine, *World Mining*, October, pp. 58-61.

Langstaff, J. 1976. Helca seismic detection system. In: *Proceedings of the 17th Rock Mechanics Symposium, Snowbird, Utah.*

Leighton, F., 1983. *Proc. of 3rd Conference on Acoustic Emmision and Microseismic Investigation.*

8

STABILITY OF STOPES BY WALL-CLOSURE MEASUREMENTS

K.S. Nagarajan, Shrikant B. Shringarputale,
N. Palani and R. Krishnamurthy

Bharat Gold Mines Ltd., Kolar, Karnataka, India

INTRODUCTION

Problems of ground control and rockbursts have been very severe in the Kolar gold mines as mining reached great depths. A systematic scientific investigation has been carried out since the 1950's to understand the phenomenon of ground movement. The problem of rockbursts is most severe in Champion Reef Mine which has reached depths over 3 km. Two of the major ore shoots of the mine, namely Glen ore shoot and North folds were severely affected by major rockbursts in November 1962 and December 1966 respectively, causing widespread damage to the workings, and were the first of their kind in the history of mining in Kolar Goldfields. The problems of re-opening and rehabilitation of these affected areas were taken up by the expert committees constituted and on their recommendations both areas were opened for mining again, adopting a new mining method known as 'Stope Drive' with concrete-fill a new method of stoping which has become a standard mining practice for the deeper workings of the mine, replacing the rill system of stoping with dry-granite walling as the medium of support practised earlier. With the resumption of mining in these areas, wall-closure measurements were carried out regularly in stopes to study the rate of closure and also the percentage consolidation of the fill, which gives an estimate of stability. Wall-closure measurements have proved very useful in assessing day-to-day stability of stopes and it has now become standard practise to utilise these measurements in areas vulnerable to rockbursts.

This paper gives details of some of the measurements carried out in the area and their practical utility in assessing day-to-day stability of stopes.

ANALYSIS OF WALL-CLOSURE MEASUREMENTS

Stope closure monitoring stations are introduced approximately 3 to 4 m behind each stope in the ladderway gaps as the face advances. Measurements are taken either daily or on alternate days, depending on the situation, for a period up to 18 to 24 months or until the stations become inaccessible. A statistical analysis of the results is done and control charts are maintained showing the "warning limit" and "action limit" to exercise better control over the stope working (Fig. 1).

$$\text{Warning limit} : \bar{X} \pm 2\sigma$$
$$\text{Action limit} : \bar{X} \pm 3\sigma$$

Where $\bar{X}$ is the grand average for the preceding 3 months and σ the standard deviation. Measurements exceeding the "action limit" and examined with care and work is suspended or men are withdrawn from the working point, if deemed necessary. Statistical analysis of results obtained from a group of stations has confirmed the existence of a significant difference 'between periods' and 'between bays' before the occurrence of some rockbursts. With more knowledge gained on closure measurements and further refinements, further analysis of data is presently being carried out and the results assessed on the significance of differences between periods only.

Wall-closure measurements in stopes have indicated a correlation between rockburst and rate of closure on many occasions. In a few instances, there was a significant increase in rate of closure varying from 2 to 16 times the normal rate for a few days prior to the rockbursts, showing the 'V' phenomenon, while in others there were no such indications. In some cases, increase in the rate of closure dropped and any subsequent increase was followed by a rockburst. It was interesting to observe that strain measurements in crosscuts also indicated an approximately 'U' phenomenon, a 'contraction' followed by an 'expansion' during periods when rockbursts occurred in the area.

With the introduction of the latest techniques in investigations of the rockburst problem in the Kolar Goldfields, viz. seismic techniques, a micro-seismic network of geophones (high frequency accelerometers) to cover an area of 100 m × 100 m between levels 98 and 103 in the south wing of the Glen ore shoot, Champion Reef mine, was established during 1983. The micro-seismic signals are monitored and analysed by a digital system and a micro-computer, to study the micro-seismic activity in the area and predict the occurrence of rockbursts. Analysis of data has been carried out to investigate the relationship between the changes in the character of the strain/stress measurements and the seismicity

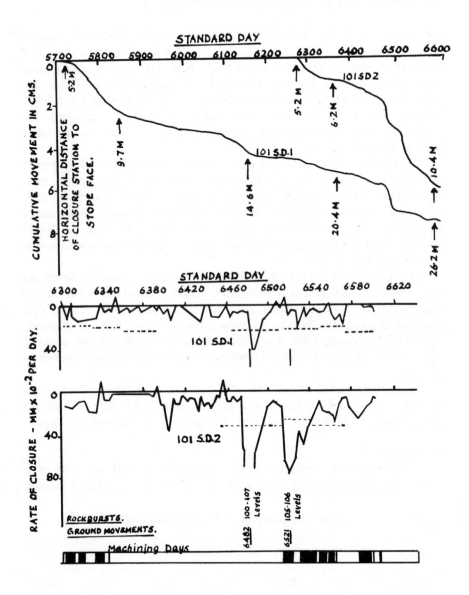

Fig. 1. Stope closure below 101 L-Northern folds area, Champion Reef Mine, B.G.M.L.

in the area during the occurrences of some rockbursts prior to and after failure.

OBSERVATIONS IN NORTHERN FOLDS AREA

The Northern folds are a part of the Champion lode system. The folded formation of the quartz reefs, which commences at the 73rd level has been developed to the 113th level, 3,200 m below the field datum. The folds have an average strike length of 60 m and in plan view are 'Z'-shaped; west limb, east limb and main reef which tends to elongate in depth where the average strike length increases to over 120 m. The general pitch of the folds is northwards. The Mysore north fault, which is a major geological feature, lies only a short distance into the hanging wall of the folds.

Extraction of the reefs was done earlier by means of a rill system of stoping with granite as waste support but after the major rockburst in 1966, which affected the area badly, the stoping method was changed to a 'stope drive' and most of the area has been stoped successfully. Wall-closure measurements were carried out in the area for assessment of the stability of the stopes and the percentage consolidation of the fill. The instrumentation system is shown in Fig. 2.

Measured closure in stopes with concrete support is found to be regular and uniform with advance of stope face and stability behind the face is reached much earlier compared to stopes supported by granite walling. The average total closure between walls in the stopes supported by concrete is approximately 10 to 13% of stoping width over a face advance of 25 m. Closure, as much as 15 to 22% of stoping width, is reached for a distance of over 50 m behind the face.

During the period of measurements, a few rockbursts/rock spitting were reported. Typical cases of significance *vis-a-vis* anomalies in the measurements against some of the bursts have been discussed in the following examples:

1) *6th July 1977*

Rock spitting was reported on this day below level 101, east limb stope drive. Wall closure recorded below level 101, on the east limb was as follows:

Rate of closure in mm × 10^{-2} per day

Bay No.	27.6.77	29.6.77	1.7.77	4.7.77	6.7.77	8.7.77
S.D.6	19.56	30.48	31.75	41.40	39.37	30.48
S.D.7	66.04	96.52	95.25	61.72	57.15	40.64

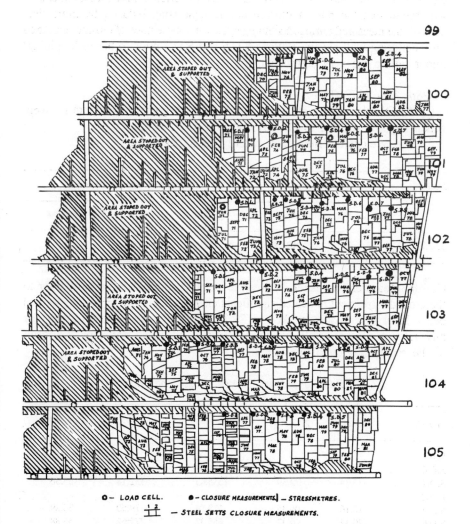

Fig. 2. Bharat Gold Mines Ltd., Champion Reef Mine Northern
folds area—East limb of fold.

The stope was under machining from day 25.6.77 and the rate of closure on day 29.6.77 was 30.48×10^{-2} mm/day at station S.D.6 (the action limit was 33.78×10^{-2} mm/day) and 96.52×10^{-2} mm/day at station S.D.7. From day 30.6.77 to day 8.7.77 the stope was under concreting but the rate of closure on day 1.7.71 was of the same order with 31.75×10^{-2} mm/day for S.D.6 and 95.25×10^{-2} mm/day for station S.D.7. Even on day 4.7.77, prior to rock spitting, the rate of closure was 41.40×10^{-2} mm/day for station S.D.6 and 61.72×10^{-2} mm/day for station S.D.7. By experience and measurements it was found that the rate of closure during concreting reduced by more than 50% of that measured during machining cycles and these rates of closure measured on monitoring wall-closure stations S.D.6 and S.D.7 prior to the rock spitting/ minor failure are very significant.

2) *28th July 1977* (Fig. 3)

The Northern folds area was shaken by a rockburst of medium intensity which slightly affected levels 101 and 102.

The Weichert seismograph record was as follows:

Date	Time		Stylus displacement in mm.		
	Hrs.	Min	N/S	E/W	Vertical
28.7.77	21	19.03	5.0	3.6	1.3

Closure measurements taken between levels 100 and 103 in the area prior to the ground movement indicated significant changes in closure at stations located below level 101. Readings taken on 29.7.77 after the rockburst indicated very high rates of closure at all monitoring stations, occurring probably as a result of the rockburst/ground movement:

Rate of closure in mm. $\times 10^{-2}$ per day

Bay No.	25.7.1977	27.7.1977	29.7.1977	1.8.1977	3.8.1977
L.101					
S.D.6	12.70	26.67	640.08	284.48	64.77
S.D.7	37.34	72.39	1535.43	551.18	35.56
L.102					
S.D.5	10.16	06.35	129.54	79.50	34.29
S.D.6	20.32	20.32	655.32	277.62	55.88
S.D.7	0	21.59	762.00	626.62	63.50

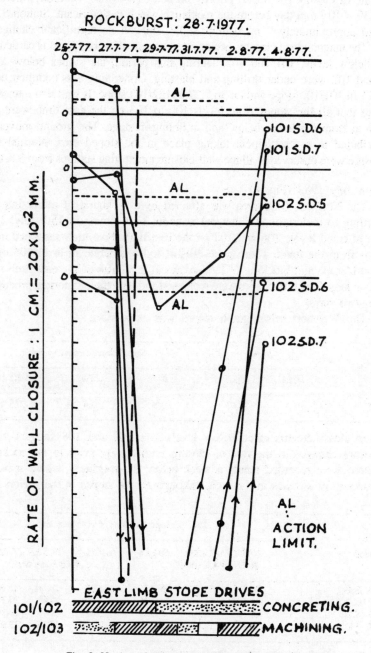

ROCKBURST: 28·7·1977.

25·7·77. 27·7·77. 29·7·77. 31·7·77. 2·8·77. 4·8·77.

AL

101 S.D.6

101 S.D.7

AL

102 S.D.5

102 S.D.6

AL

102 S.D.7

AL
ACTION
LIMIT.

RATE OF WALL CLOSURE : 1 CM.= 20×10⁻² MM.

EAST LIMB STOPE DRIVES

101/102 —— CONCRETING.

102/103 —— MACHINING.

Fig. 3. Northern folds area—Champion Reef Mine.

The rate of closure for S.D.7, priorily around 37.34×10^{-2} mm/day, increased to 72.39×10^{-2} mm/day before the rockburst, which is significant. Station S.D.6, located approximately 9 m, behind S.D.7, showed no significant change.

The magnitude of closures occurring as a result of the rockburst is particularly high below levels 101 and 102. During this period, the stopes below levels 101 and 102 were under drilling and blasting. Concreting was commenced on 29.7.77 in 101/102 stope and on 30.7.77 in 102/103 stope. It was also interesting to note that all the stopes from levels 100 to 103 on the east limb were only 4 to 6 m from the levels below and in holing-in stage. The ground movement is attributed to a resettlement taking place in the stoped area. Normal rates of closure were observed at all the wall-closure monitoring stations from 3.8.1977.

3) *15th July 1983* (Fig. 4)

The Northern folds area was affected by a rockburst of minor intensity (according to underground damages) at about 1740 hours on 15th July, 1983, which affected levels 106 and 108 on the east limb. Five men sustained minor injuries from the burst. There were falls of loose concrete in level 106 of the east limb stope and blocking of ladderways at level 108 of the east limb stope by fall of loose concrete. Stopes were reclaimed and brought into normal production within two days.

The Weichert seismograph record was as follows:

Date	Time		Stylus displacement in mm.		
	Hrs.	Min.	N/S	E/W	Vertical
15.7.83	17	39.67	0.8	0.4	0.5

Closure measurements taken below levels 106, 107 and 108 showed highly significant changes in the rate of closure immediately prior to the rockburst and these were recorded nearly a week before the rockburst, which indicated the build-up of stresses in the area, although all the stopes in the region were

Bay No. East Limb	Rate of closure in mm $\times 10^{-2}$ per day					
	9.7.83	12.7.83*	15.7.83	16.7.83	19.7.83	21.7.83
	.... Before Rockburst......			...........After Rockburst.......		
L. 106 S.D.2	34.29	51.56	36.83	64.77	27.94	12.70
L. 107 S.D.2	25.40	71.88	36.83	0	9.40	20.32
L. 108 S.D.1	12.70	53.34	27.94	5.08	9.40	10.16
L. 108 S.D.2	16.51	72.90	29.21	10.16	17.02	13.97

*All rates of closure per day crossed the 'action limit'.

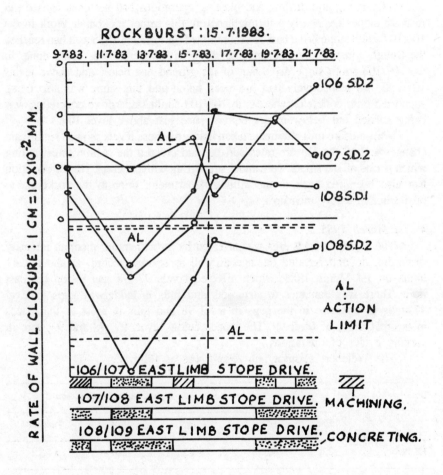

Fig. 4. Northern folds area—Champion Reef Mine.

on the supporting cycle with concrete-fill being carried out. As a result of the rockburst, there were significantly high rates of closure and equilibrium attained within a few days. The following Table gives details of closure prior to and after the minor rockburst.

During this period all the stopes, viz. 106/107 south stope drive, 107/108 south stope drive and 108/109 south stope drive, all on the east limb, were under concreting and drilling and blasting operations had not been carried out in these stopes for nearly a fortnight before this minor rockburst. Work in the 106/107 south stope drive had been completed during June 1983 as it had reached the trough winze and supporting the stope with concrete was being done. In the 107/108 south stope drive part of the ground just below and above levels 107 and 108 respectively had just been taken and this stope was also being supported with concrete, whereas in 108/109 south stope drive concreting was being carried out below in the bottom slice just above level 109.

Wall-closure measurements taken below all these levels showed significant changes and all the closure rates per day had crossed the 'action limit' during which period all the stopes were under supporting with concrete. This information had also been fed back to the mining department, prior to the rockburst to take suitable safety measures.

4) *1st March 1985*

The Northern folds area was affected by a rockburst of medium intensity (according to underground damage as well as seismic record) at about 2241 hours on 1st March 1985, which affected levels 97, 98 and 99 on the east limb. There was damages to steel sets and falls in ladderway gaps in level 97, falls of concrete in the gaps in level 98 and falls in most of the places in the reef drive of level 99. The stopes below levels 97, 98 and 99 were in varying cycles of operations.

The Weichert seismograph record was as follows:

Date	Time		Stylus displacement in mm		
	Hrs.	Min	N/S	E/W	Vertical
1.3.1985	22	40.00	8.2	5.0	5.2

Seismic network recording details were as follows:
Date: 1.3.1985 *Time*: 22 hrs. 40.58 min. Amplitude: 110 mm. *Duration of tremor*: 120 seconds. *Computed co-ordinates*: Latitude: 19258'S. Departure: 4647'W. B.F.D.: 9713 Ft.

Closure measurements taken below levels 97 and 98 showed high rates of closure prior to the burst, which took place nearly 12 hrs. after measurements taken the morning of the same day. The 97/98 south stope drive was on drilling and blasting operations and the other two stopes (98/99 south stope drive and 99/100 south stope drive) were under miscellaneous operations other than drilling and blasting or supporting operations. As a result of the rockburst, there were abnormally high rates of closure and the area attained equilibrium a few days later.

The following Table gives details of closure prior to and after the rockburst:

| Bay No. | Rate of closure in mm $\times$ 10^{-2} per day | | | | | |
| | 25.2.85 | 27.2.85 | 1.3.85 | 4.3.85 | 6.3.85 | 11.3.85 |
East limb:						
L. 97 S.D.3	20.32	7.62	77.47	Out of	541.83	20.82
L. 98 S.D.2	0	2.54	59.69	range	Out of range	

During this period, it was observed that whereas the 97/98 south stope drive was on drilling and blasting operations, the 98/99 and 99/100 south stope drives were idle.

Wall-closure measurements taken on 1.3.85 below levels 97 and 98 before the medium rockburst showed very high rates of closure, which were approximately 10 and 20 times the closure rates recorded two days earlier on 27.2.1985. It was also very interesting to note that when wall-closure measurements were being monitored on 1.3.85, rock spitting was also noticed in the area. Rock spitting is quite commonly noted, whenever anomalies in wall closures are observed at the measuring stations.

The experiences from Northern folds area may now be summarised. During the period of stope-closure measurements from 1972 to 1986 in the area between levels 97 and 108 on the east limb, 33 rockbursts/rock spitting incidents were reported, of which, 10 were of major/medium intensity and the rest minor. Of these, on 12 occasions there was a significant increase in rate of closure prior to rockburst/rock spitting in stopes and for the remaining 21 rockbursts/ resettlements no such indication prior to the occurrence. But on every occasion abnormal closures were measured after the occurrence of rockbursts or ground movements, occurring probably as a result of the same.

In more than 200 instances, the mine manager/mining engineers have been informed when anomalies were observed in the measurements in the area and adequate precautions such as stoppage of blasting for a day or two and accelerating support operations etc. were taken to stabilise the area.

OBSERVATIONS IN GLEN ORE SHOOT

The Glen ore shoot, a large ore body on the Champion lode system in Champion Reef mine extends from level 68 to the bottom of the mine at level 110 and is about 400 m long on the strike. The reef is about 1 m wide, dipping 84° westward. A major fault known as "Mysore north fault" striking NNW converges on the ore shoot especially on the north wing. Approach to the stoping area is by two major vertical shafts—one on the north side known as Heathcote shaft and the other on the south side known as Osborne shaft—and by crosscuts and footwall drives.

The extraction system earlier was by rill system of stoping with granite as waste support in both the south wing and north wing of Glen ore shoot, but after the major area rockburst in 1962/1963, which affected the area badly the method of stoping was switched to "stope drive" system with concrete as the medium of support. Approaches to the ore shoots were reclaimed, the rill promontories shaped and a longwall system of stoping adopted between levels 98 and 103, first on the south wing of Glen ore shoot and next on the north wing. The sequence of stoping is maintained in such a way that the stope face below the level is always kept one stope length in advance of the stope face immediately above the level. To date, considerable stoping has already been carried out successfully. Wall-closure measurements are conducted in the area for assessment of stope stability and percentage consolidation of fill. The instrumentation layout is shown in Fig. 5.

The measured closure in stopes with concrete support is found to be more regular as observed, for example, in the northern folds area, and is uniform with the advance of stope face. Stability behind the face is reached much earlier compared to stopes supported by granite walling. The average total closure between walls in stopes supported by concrete, is approximately 10 to 15% of stoping width over a face advance of 25 m. Closure of as much as 15 to 23% of stoping width is reached for a distance of over 50 m behind the face.

During the period of measurements, a few rockbursts/rock spitting were reported. Typical cases of significance of anomalies in the measurements for some of these rockbursts are discussed in the following examples. In addition, these anomalies have also been examined vis-a-vis the micro-seismic activity being monitored in the south wing of Glen ore shoot since 1985.

1) *9th February 1978*

The 98/99 south stope drive was under machining during the afternoon shift on 9.2.78 when spitting was observed in the stope. Stope work was therefore suspended for the day in 98/99 south stope drive as well as in 99/100 south stope drive. There was no Weichert seismograph record for this event.

Closure measurements taken in the area on the same day prior to rock spitting indicated significant changes in rates of closure below levels 98 and 99 as follows:

The high rates of closure recorded below levels 98 and 99 on 8.2.78 were discussed with the section mining engineer. During this period all the stopes were in the position of holing through to the levels below and were under a drilling and blasting cycle except for stoping below levels 101 and 102, which were under concreting after completion of drilling and blasting.

It has been the usual experience in the Osborne shaft area that rock

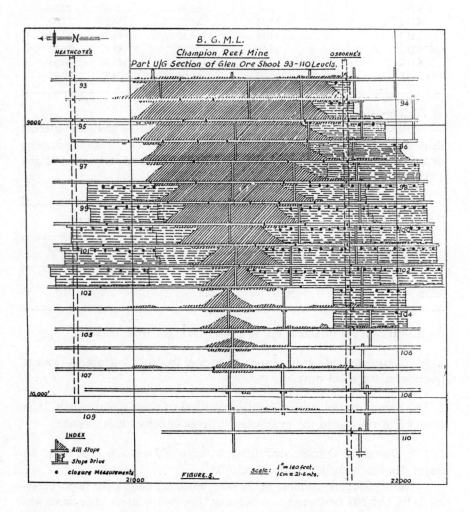

Fig. 5. Stope drive system—Glen ore shoot Champion Reef Mine.

spitting/minor ground movements occur during periods when the first cut immediately below the level and/or the last holing cut through to the level below is carried out.

Bay No.	Rate of closure in $mm \times 10^{-2}$ per day			
	6.2.78	*8.2.78*	*10.2.78*	*13.2.78*
98 S.D.1	9.40	53.34	25.40	22.10
S.D.2	15.24	82.55	34.29	17.02
S.D.3	10.16	114.30	48.26	22.10
99 S.D.5	10.16	123.19	83.82	36.32
S.D.6	29.72	134.62	143.51	40.64
100 S.D.6	22.10	66.04	109.22	27.94

2) *7th/8th July 1978*

The Osborne shaft area in the southern wing of Glen ore shoot was affected by medium rockbursts which occurred on 7th and 8th July 1978 causing damage to levels 97/98 in the stoping region and the Osborne shaft between 102 and 103 level plats. The Weichert record was as follows:

Date	Time		Stylus of displacement in mm		
	Hrs.	*Min.*	*N/S*	*E/W*	*Vertical*
7.7.78	14	59.07	4.5	2.4	3.0
8.7.78	05	06.80	5.0	1.3	1.5

As a result of the rockburst the following damages were noticed:

L. 97: Damage to laggings support at the back of the No. 1 companion crosscut junction in the south drive.

L. 98: Damage to steel sets and lagging poles from a point approximately 25 m from 97/98 rill end to the reclamation face at Osborne shaft crosscut.

L. 102: Damage to Osborne shaft lining between 102 and 103 level plats, the major damage being just below level 102.

During this period drilling and blasting was being done in the stopes below levels 98 and 100 subsequent to concreting. The 99/100 south stope drive was idle after a blast on 6.7.78. The stopes below levels 101 and 102 were under a drilling and blasting cycle after concreting had been completed. The 98/99 and 99/100 south stope drive faces were lying opposite the shaft and the south stope faces below levels 100 to 102 had crossed the shaft.

Heavy rock spitting was reported in the morning shift on 6.7.78 from the south area at levels 100, 101 and 102.

Closure measurements taken in the area were as follows:

Bay No.	Rate of closure in mm × 10^{-2} per day					
	30.6.78	3.7.78	5.7.78	7.7.78	10.7.78	11.7.78
				Rockburst		
98 S.D.1	20.32	12.70	7.62	38.10	512.32	25.40
S.D.3	12.70	22.86	11.43	1.27	583.44	33.02
S.D.4	0	153.16	38.10	68.58	Not accessible	
99 S.D.5	13.97	6.86	16.51	52.07	381.00	38.10
S.D.6	17.78	12.70	19.05	120.65	336.04	33.02
S.D.7	0	153.16	82.55	157.48	233.68	43.18
100 S.D.6	13.97	21.08	19.05	107.95	196.34	38.10
S.D.7	33.02	34.80	21.59	158.75	155.70	15.24
101 S.D.7	12.70	12.70	19.05		Fall of loose rock	
S.D.8	12.70	8.38	19.05	154.94	121.16	10.16
S.D.9	17.78	28.70	22.86	240.03	113.54	7.62
102 S.D.10	—	16.00	—	95.25	99.82	10.16
S.D.11	—	17.02	—	171.45	83.06	10.16
S.D.12	12.70	17.02	25.40	266.70	75.44	12.70
S.D.13	—	—	0	416.56	55.12	12.70

Closure measurements taken between levels 98 and 103 on the morning of 7th July 1978 showed significant increases prior to the rockburst, which occurred at 1450 hours on the same day, resulting in damages to levels 97 and 98 of Osborne shaft.

3) *19th November, 1980*

Stray falls of ground were observed in the south wing of Glen ore shoot between levels 98 and 103, which were attributed to a suspected minor rockburst on 19.11.80. All the south stope drives were found unaffected by the suspected rockburst.

The Weichert seismograph record was as follows:

Date	Time		Stylus displacement in mm.		
	Hrs.	Min	N/S	E/W	Vertical
19.11.80	19	36.45	10.7	4.0	3.6
	19	46.22	0.4	0.9	0.4

During November 1980, the south stope drives between 100/101 and 102/103 were holing through to the respective levels below. The 101/102 south stope drive was being worked at the 'top cut'. Extraction of sill below level 98 was completed just at that time. During this period, all the stopes between levels 98 and 103 were under drilling and blasting operations.

Measurements taken in the south wing the morning of the 17th and also the 19th before the rockburst, showed a high rate of closures and this was discussed with the mining engineers. Heavy rock spitting was also reported from levels 98, 99 and 100 before the rockburst. Rates of closure measured before and after the rockburst are given below:

Bay No.	Rate of closure per day $\times$ 10^{-2} mm						
	14.11.80	17.11.80	19.11.80	21.11.80	24.11.80	28.11.80	1.12.80
			Before Rockburst				
L. 98 S.D.6	7.62	40.64	99.06	109.22	42.16	22.86	12.70
L. 99 S.D.8	30.48	27.18	86.36	233.68	58.42	43.18	16.00
S.D.9	+ 3.81	38.10	93.98	93.98	45.72	39.37	16.00
L. 100 S.D.8	39.37	23.62	24.13	—	103.63	45.72	12.70
S.D.9	7.62	43.94	101.60	245.11	69.34	56.39	25.40
L. 101 S.D.11	8.89	25.40	63.50	147.32	49.78	41.15	21.84
L. 102 S.D.15	20.32	16.00	46.99	113.03	45.72	27.18	20.32

Closure measurements on the 21st after the rockburst showed high rates of closure at all the closure stations. The closure became normal from 1.12.80. It is interesting to note that ground movements/resettlements usually occur when stope faces reach the stage of holing through to the level below.

This rockburst was recorded by the seismic network and the computed foci agrees with the location of the damages reported.

4) 1st/2nd December 1985 (Fig. 6)

The southern wing of Glen ore shoot was affected by a medium rockburst which occurred on 1st/2nd December 1985 causing stray falls of rock and concrete support in levels 99, 100, 101 and 102, which included shaft crosscuts and reef drives. There were stray falls observed in the footwall drives also.

The bursts were recorded by the seismic network as two rockbursts, one at 1857 hrs on 1.12.85 and the other at 0152 hrs, on 2.12.85. The one that took place at 1857 hrs on 1.12.85 was of a major magnitude with an amplitude of 115 mm and a duration of 60 seconds, the other 95 mm amplitude and duration 30 seconds.

Closure measurements taken in the area during the period were as follows:

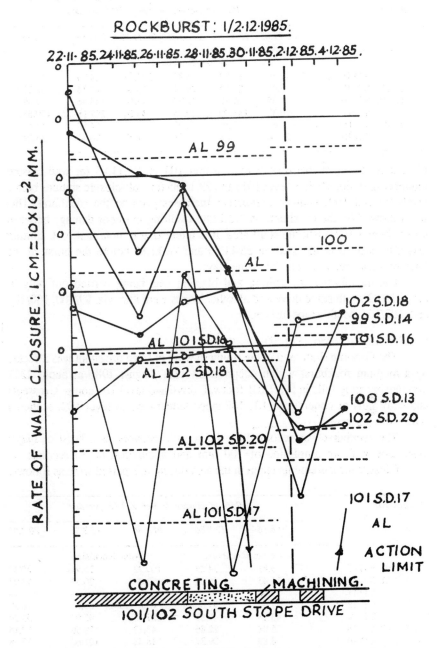

Fig. 6. Glen ore shoot—Champion Reef Mine. Rockburst event on
1/2nd December, 1985.

Bay No.	Rate of closure in mm $\times$ 10^{-2} per day					
	22.11.85	25.11.85	27.11.85	29.11.85	2.12.85	4.12.85
	Before Rockburst					
L. 99 S.D.14	12.70	83.82	55.88	223.52	114.30	111.76
L. 100 S.D.13	5.08	24.64	29.21	66.04	142.24	128.27
L. 101 S.D.16	6.35	61.72	12.70	43.18	140.46	72.39
S.D.17	20.32	143.00	19.05	48.26	297 18	120.65
L 102 S.D.18	7.62	19.56	5.08	0	54.10	11.43
S.D.20	27.94	5.08	3.81	1.27	36.32	34.29

Measurements taken between levels 98 and 103 on 25.11.85 (at wall-closure monitoring stations below level 101) and 29.11.85 (at wall-closure stations below levels 99 and 100) showed significant increases prior to the rockburst. This was followed by the rockburst on 1/2.12.85 resulting in minor damage between levels 99 and 102, Osborne shaft area. It is interesting to note that rock spitting was also noticed in the area on 25.11.85 and 28.11.85 before the burst, at the time of measurements on the same days.

During this period, drilling and blasting was being carried out only in the 101/102 south stope drive and all other stopes were idle, viz. 99/100, 100/101 and 102/103 south stope drives.

5) *13th August 1986* (Fig. 7)

The Osborne shaft area in the southern wing of Glen ore shoot was affected by a medium rockburst which occurred on 13th August 1986 at about 2351 hrs, causing stray falls in level 101 footwall drive and level 101 comp. Crosscuts and stray falls of concrete in 101/102 stope ladderway. In level 102, scattered falls were observed.

The rockburst was recorded by the seismic network as a burst of major magnitude with an amplitude of 100 mm and a duration of 50 seconds.

Closure measurements taken in the area during this period were as follows:

Bay No.	Rate of closure in mm $\times$ 10^{-2} per day				
	10.8.86	12.8.86	14.8.86	18.8.86	20.8.86
	Before Rockburst		After Rockburst		
L. 100 S.D.13	3.81	34.29	109.22	25.91	20.32
L. 101 S.D.16	11.43	59.69	53.34	20.32	13.97
S.D.17	25.40	71.12	—	—	
S.D.18	7.62	99.06	Filled with rubble	89.66	20.32
S.D.19	43.18	147.32	227.33	85.09	53.34
L. 102 S.D.18	5.08	22.86	43.18	19.56	10.16
S.D.20	8.89	39.37	138.43	101.60	17.78
S.D.21	24.13	74.93	375.92	44.96	1.27

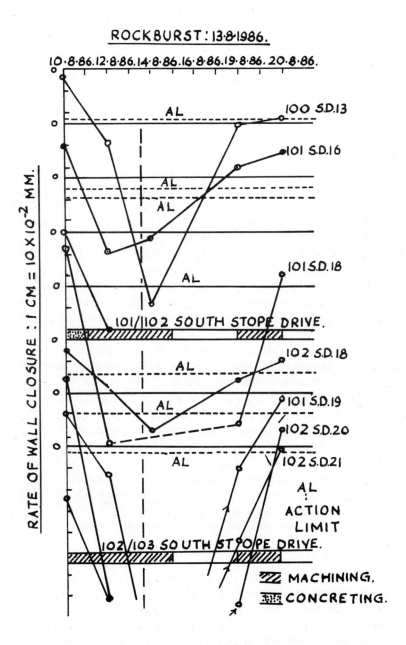

Fig. 7. Glen ore shoot—Champion Reef Mine. Location of
a rockburst on 13.8.1986.

Closure measurements taken below levels 101 and 102 on 12.8.86 showed highly significant changes at all closure monitoring stations prior to the rockburst on 13.8.86 and closures were of the order of nearly 3 to 10 times the closures recorded on 10.8.86. All these rates of closure exceeded the 'action limit'. The rates of closure further increased as a result of the rockburst and recorded on 14.8.86 as shown in the Table. The area attained equilibrium and the rate of closure became normal from 20.8.86.

During this period, the 100/101 south stope drive was idle and the other two stopes viz. 101/102 and 102/103 south stope drives were on a drilling and blasting cycle after completing concrete support.

OBSERVATIONS CAN NOW BE SYNTHESISED

During the period of measurements from 1972 to 1986, 66 rockbursts/rock spitting were reported, of which 7 were major/medium in intensity and the rest minor. All the incidents occurred in the area between levels 98 and 103, in the south wing of Glen ore shoot, where stope-closure measurements were being conducted. Of these, on 26 occasions there was a significant increase in the rate of closure prior to rockburst/rock spitting in the stopes and for the remaining 40 rockbursts/resettlements no such indications prior to the occurrences were noticed. But on every occasion abnormal closures were measured after the occurrence of rockbursts or ground movements, probably as a result of the same.

An attempt was made to correlate the micro-seismic activity recorded between levels 98 and 103 of south section with closure measurements carried out in the same area. In a few instances it was found that the increase in microseismic activity coincided with the increase in rate of closure but lagging in time noted in others. There was also a rise followed by a decrease in seismic activity versus an increase in rate of closure prior to some rockbursts/rock spitting that occurred in the area. However, details of the trend are being examined further for better appreciation of the data collected prior to rockbursts.

In more than 300 instances, mine managers/mining engineers were informed when anomalies were observed in wall-closure measurements carried out in the area and adequate precautions such as stoppage of blasting for a day or two and accelerating support operations etc. were taken to stabilise the area.

CONCLUSIONS

Analysis of closure measurements between stope walls taken in selected areas has provided valuable information on the consolidation of fill and general strata movement. The following are considered of great interest.

1) Measured closure consisted of a somewhat gradual closure due to mining and large sudden closures due to rockbursts.

2) Rate of closure at a point was usually high near the face and gradually decreased as the face moved away.

3) The magnitude of maximum closure measured varied from point to point and also depended on mining methods and support system adopted, stress conditions and behaviour of wall rocks.

4) An increase in rate of closure was associated with machining operations carried out in the stope and also rockburst.

5) Rates of wall closure in stopes revealed significant changes or anomalies in the trend of closure during periods of rockbursts; in many instances there was an increase in rate of closure or an increase followed by a fall in rate of closure prior to the occurrence of rockbursts/rock spitting, and on other occasions neither of these anomalies was observed.

6) A close relationship was noted between wall closure and micro-seismic activity of the area in the context of occurrence of rockbursts.

7) Wall-closure measurements have proved very useful in assessing day-to-day stability of stopes. It has now become a standard practise to introduce these measurements in all areas vulnerable to rockbursts.

Hopefully, closure measurements along with micro-seismic data may eventually lead to the prediction of rockbursts.

ACKNOWLEDGEMENT

The authors are grateful to the Chairman and the Managing Director of Bharat Gold Mines Limited for permission to present this paper.

REFERENCES

1. BGML Research and development unit special reports (unpublished).
2. Blake, W. 1971. Rockburst research at the Galena mine, *USBM TPR* No. 39, August 1971, 22 pp.
3. Blake, W. and F.W. Leighton. 1970. Recent development and application of the Micro-seismic method in deep mines, pp. 429-443. In: *Rock Mechanics: Theory and Practice* edited by W.H. Somerton. A.I.M.E., New York.
4. Hardy Jr., H.R. and F.W. Leighton. 1984. Proceedings of the Third Conference on 'Acoustic Emission/Micro-seismic Activity in Geological Structures and Materials. Trans. Tech. Publications.
5. Krishnamurthy, R. and K.S. Nagarajan. 1976. Strata control measurements in stopes. Golden Jubilee Symposium. B.H.U. 1976.
6. Krishnamurthy, R., K.S. Nagarajan and N. Sethumadhavan. 1980. A study of stope-closure measurements in deep levels of Champion reef mine. Seminar on Recent Trends in Gold-Mining Practise, Bharat Gold Mines Limited, Dec. 1980.
7. Krishnamurthy, R. and K.S. Nagarajan. 1981. Rockbursts in Kolar Goldfields; Rock Mechanics. Proc. Indo-German Workshop on Rock Mechanics, Hyderabad, Oct. 1981.
8. Krishnamurthy, R. and P.D. Gupta. 1983. Rock mechanics studies on the problem of ground control and rockbursts in the Kolar Goldfields. *Symp. on Rockbursts: Prediction and Control, October 1983*. IMM, London.

9. Miller, E. Notes on rock mechanics research in K.G.F., *K.G.F. Min. Metall. Soc. Bull.*, vol. 95, pp. 23-83.
10. Potts, E.L.J., E. Miller and R. Krishnamurthy. 1965. Rockburst investigations in the Kolar Goldfields, South India. 7th Symp. of International Bureau of Rock Mechanics, Leipzig, Nov. 1965.
11. Special Committee Report on Occurrence of Rockbursts in the Kolar Goldfields, 1955.
12. Taylor, J.T.M. 1963. Research on ground control and rockbursts on the Kolar Goldfields. India, *Trans. Inst. Min. Metall. London* (1962-63) vol. 72, pp. 317-338.

COAL MINE BUMPS—CASE HISTORIES

9

BUMPS AND ROCKBURSTS IN INDIAN COAL MINES—AN OVERVIEW

A.K. Ghose

Indian School of Mines, Dhanbad 826 004

INTRODUCTION

Rockbursts and bumps in coal mines have been recognised as a major hazard for many years. The problem dimension of late, however, has exacerbated due to increased working depths, higher intensity of exploitation and more difficult geological environments. Singh *et al.* [1] have reported that between 1944 and 1964, there were some 94 bump occurrences causing 136 fatal accidents in Indian coal mines. According to their analysis, pillar extraction accounted for 85% of the total accidents, including 58.4% on the faces and 26.6% in split galleries; the longwall face at Parbelia (worked with stowing) alone contributed 9.5% of the bumps.

More recent data, summarising the decadal experience from 1976 to 1986, on the incidence of bumps in Indian coal mines are presented in Table 1.

It must be conceded, however, that in the absence of an accurate description for a bump occurrence, the data could be somewhat misleading. Statistics also do not reflect the overall magnitude of bump occurrences, recording only those involving fatal or serious accidents.

In the coming years, with the envisaged expansion in coal production, the average depth of extraction is expected to rise significantly and some of the seams in Raniganj Coalfield may reach extraction levels of over 700 m. This paper seeks therefore to appraise the problems of coal mine bumps in

Table 1. Fatal and serious accidents in coal mines due to bumps

Year	No.	No. of persons killed	No. of accidents	No. of persons seriously injured
1976	2	2	12	12
1977	–	–	8	8
1978	–	–	8	8
1979	–	–	11	11
1980	–	–	7	12
1981	–	–	5	5
1982	–	–	1	1
1983	1	1	2	3
1984	1	1	1	1
1985	1	1	2	2
1986	1	–	1	1

Indian coal industry, outlining the current research efforts. An overall strategy for the prognosis and for the control of coal mine bumps is also outlined.

COAL MINE BUMPS—A PERSPECTIVE

The semantic problems of defining a coal mine bump notwithstanding, it is now common to use the term bumps and rockbursts synonymously. If one could provide a more practical definition, it is an instantaneous explosive destruction of coal in the mass without significant gas emission with concomitant impact on men, materials and structures. Ejection of coal, as distinct from free fall due to gravity, is a necessary criterion for defining a bump. Destructive rock failure in the working area would also come under this category. The causative factors of bumps have been analysed from several standpoints, but a bump is essentially a manifestation of excessive stress, accompanied by seismic energy release, which results in pillar or abutment collapse, rotational collapse of cantilevered beds over a longwall etc.

A host of contributory factors can lead to the causation of a bump which include *inter alia* the following:

— ability of the coal seam to store and release strain energy;
— ability of roof and floor rocks within a short distance of the seam to store and release seismic energy, namely, strong roof and floor;
— geological factors such as depth, presence of faults, dykes etc.;
— mountainous terrain; and
— mining configuration leading to high stress concentration.

HAZARDS DUE TO BUMPS IN INDIAN COAL MINES—AN OVERVIEW

As the average depth of underground working in Indian coal mines is relatively low (of the order of 180–200 m), the problem dimension is presently not acute. At specific locales, where depths are greater, such as in Raniganj Coalfield, in the workings of Dishergarh seam at Chinakuri and Parbelia Colliery and of Koithi and Poniati workings at Girimint Colliery, there are potent problems of bumps. In fact, the percentage of reserves in Raniganj Coalfield at depths exceeding 300 m aggregate to over 32%.

The occurrence of bumps in Raniganj Coalfield date from around the 1920s, the greatest intensity of these being recorded in and around Parbelia, Dhemo Main and Chinakuri in the workings of Dishergarh seam at depths of near 500 m. The bump experiences of the 1940s called for some radical changes in mining strategy, with the adoption of longwall stowing at Parbelia, pillar extraction with stowing using the slicing method and avoidance of pillar splitting. These have been reflected in improved statistics of bump occurrences. The causes of coal mine bumps and the combative measures against their occurrence have been examined by a host of workers, including amongst others Barraclough [2], Singh [3, 4], and Singh et al. [1]. In general, the avoidance of stress-inducing configurations and adoption of stowing longwalls led to improved control of bumps.

CURRENT RESEARCH ON COAL MINE BUMPS

A major initiative to study the problems of rockbursts in deep coal mines was undertaken by Indian School of Mines with support from an S & T Grant from the Department of Coal, Government of India between 1978 and 1986. Under this project, the focus of research effort was at Chinakuri 1 & 2 pit Colliery where a thick seam longwall caving panel was proposed to be worked in Dishergarh seam. Initial studies concentrated on the identification of bump proneness of the seam and enclosing rocks. Creep characteristics of the Dishergarh seam and its W_{ET} index were established, as also the effect of water on W_{ET} index [5, 6]. As a tactical measure, 'drilling yield' tests were carried out in the mine to help define the zones of abutment stresses. Studies carried out in the vicinity of AS-8 stowing longwall panel defined the zone of abutment about 2 m beyond the coal front.

As a part of this programme, seismic investigations were also initiated at the locale of Chinakuri 1 and 2 pits Colliery by establishing a rockburst monitoring network. This comprised a Raccal 14-track Geostore field recorder connected to a tripartite network of Wilmore MK IIA Seismometers [7]. The installation was primarily designed to monitor a caving longwall panel, which has been overly delayed because of apprehensions about the problems of bumps in Dishergarh seam. Between 1977 and 1979, an Indo-USSR Study Group investigated the problems of extraction with longwall caving and concluded that for effective control of longwall caving, a minimum support resistance of 120 t/m² would be required with destressing blasts in the roof. In the absence of destressing blasting, the support resistance would be of the order of 320 t/m², which was beyond the reach of the then available supports [8].

A geotechnical log of the roof strata at the site of the longwall locale to assess cavability has been undertaken. The W_{ET} index of the beds has also been evaluated, which shows that the longwall face will have to contend with problems of rockburst by adopting appropriate prophylatic measures, such as water infusion, hydrofracturing and 'drilling yield' tests.

During 1986, an investigation of the problems of bumps in pillar extraction at Koithi seam of Girimint Colliery in Raniganj Coalfield was undertaken [9]. At the mining locale, the 4.8 m thick Koithi seam is being extracted by caving with pillars 45 m × 45 m in size. The manifestation of abnormal ground pressures in splitting the pillars and large-scale displacement of roof coal in one split gallery called for this specific investigation. The geotechnical evaluation of roof cores up to 20 m from the seam roof revealed the existence of rockburst-prone strata with W_{ET} values exceeding 10. Drilling yield tests were carried out in the panel and Fig. 1 shows a plot of drilling yield in litres per metre against the depth of holes drilled. For a comparative evaluation, the results from drilling yield tests at Chinakuri 3 pit are shown in Fig. 2. By and large, the liability of the seam to bursting at the sites of measurement was established as low, if not marginal. The existence of roof strata with a high W_{ET} index suggested that the fracture of strata could lead to seismic events and the adoption of appropriate prophylactic measures was therefore suggested.

Although data on bump occurrences in Indian coal mines are very limited, rock mass classification for evaluating rockburst hazard at any operational site is being attempted. The first step is the classification of seams on W_{ET} index, with attention then focussed on seams with W_{ET} values greater than 2. A rockburst hazard classification is being evolved considering **interval parameters** of W_{ET}, compressive strength, depth and **nominal parameters** of structure, mining geometry etc.

RECOMMENDATIONS FOR RESEARCH ON COAL MINE BUMPS

In the perspective of Indian coal industry, the problems of rockbursts are still not of significant dimension, except in localised areas in Raniganj Coalfield. The extension in depth of workings, and the need to exploit deeper coal seam reserves call for new initiatives in coal mine bumps research so that the overall problem is kept under continuing surveillance. Table 2 gives a list of mines in Raniganj Coalfield where working depths will exceed 300 m.

R & D efforts need, therefore, to be focussed on the following problem

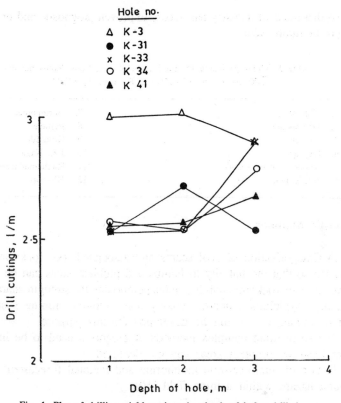

Fig. 1. Plot of drilling yield against the depth of holes drilled.

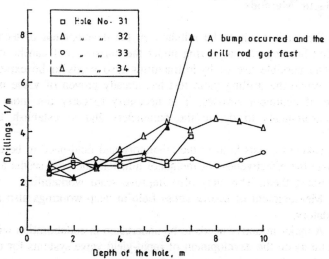

Fig. 2. Results from drilling yield tests.

areas so that the coal industry has access to proven prognostic and prophylactic methods in future years.

Table 2. Mines projected for production of coal from depths beyond 300 metres to 1999-2000 in Raniganj Coalfield

1.	Chinakuri	7.	Barmundia
2.	Dhemonain	8.	Parbelia
3.	Ningah	9.	Kottadih
4.	Satgram	10.	J.K. Nagar
5.	Sripur (Taltore seam)	11.	Radhamadhabpur
6.	Amritnagar	12.	Bhanora

Prognostic Methods

1) Categorization of coal seams and associated roof rocks in terms of W_{ET} index so that the liability to bumps and problem areas can be identified. A more complex rockburst liability index comprising the strength characteristics, W_{ET} index, exploitation depth, geology and tectonics, mining system, areal extent of mining etc., could be developed for this purpose.

2) Use of more complex methods of prognosis need to be investigated so that localised problem areas can be identified.

3) Use of micro-seismic monitoring and regional forecast of degree of rockburst hazard within a mine field [10].

Prophylactic Methods

1) Research needs to be initiated on the development of techniques of destressing blasting so that caving under the massive roof can be carried out safely. The possible use of hydrofracturing also needs to be explored.

2) While the drilling yield test has already proven of value in defining the zone of abutment stresses, it is necessary to carry out more extensive field measurements to define the parameters and to establish their variability.

3) Likewise, tests both in the laboratory and field need to be undertaken to evaluate the effectiveness of methanol infusion into coal seams as a means of destressing them. This may also improve seam workability.

4) Measurement of *in-situ* stress field in deep workings also appears to be mandatory.

5) A major initiative needs to be undertaken in collaboration with support manufacturers on the development of rapid-yield valve systems for use at coal faces liable to bumps.

ACKNOWLEDGEMENT

Grateful acknowledgement is made to the Department of Coal, Government of India, for support of S & T Project on "Rockbursts in Deep Coal Mines" under which many of the studies have been undertaken.

REFERENCES

1. Singh, T.N., M.A. Rafique and B. Singh. 1970. Bumps in coal mines. *Trans. Mining, Geological and Metallurgical Institute of India*, vol. 67.
2. Barraclough, L.J. 1950. Hydraulic stowing in India, *Trans. Mining, Geological and Metallurgical Institute of India*, vol. 45, p. 146.
3. Singh, R.D. 1976. Planning and exploitation of high production faces in deep coal seams liable to bumps. *Proc. IXth World Mining Congress, Dusseldorf,* May, 1976.
4. Singh, R.D. 1964. A modified method of extracting pillars in a thick seam in Raniganj Coalfield liable to bumps. *Proc. International Symposium on Working Thick Coal Seams,* Dhanbad, 1964. Paper No. 9.
5. & 6. Unpublished Research Report, Indian School of Mines, Dhanbad, 1979, 1980.
7. Ghose, A.K. and R.K.S. Chouhan. 1982. Bumps in Indian coal mines: A programme of investigation of Chinakuri 1 & 2 Pit Colliery. *Proc. Indo-Polish Symposium on Geophysical Applications in Mining,* Dhanbad, 1982.
8. Ghose, A.K. 1987. Design of longwall supports under massive Coal Measure rocks. *Proc. Int. Symposium on Safety in Mines Research.* Beijing. Nov., 1987.
9. Ghose, A.K. 1986. Evaluation of Bump Proneness of Koithi Seam at Girimint Colliery—An Investigation. S & T Project Report No. 4017/MIN/86/1, Dec. 1986.
10. Petukhov, I.M. 1987. Forecasting and combating rockbursts. Recent Developments. *Proc. International Society of Rock Mechanics,* Montreal, 1987.

10

STUDIES ON BUMP PRONENESS OF DISHERGARH SEAM AT BARMONDIA COLLIERY

P.R. Sheorey and B. Singh

Central Mining Research Station, Dhanbad 826 001

INTRODUCTION

Some coal seams in India, particularly Dishergarh from Raniganj Coalfield, have been reported to have caused coal bumps in the past. Such reports, however, have never been recorded in detail as case studies nor received attention as investigation problems. One such case, which occurred in Dishergarh seam at Barmondia Colliery in 1984, resulting in one fatality, could be studied in details by CMRS. The attention of CMRS has been drawn to such cases since this occurrence in 1984 and another case is at present under investigation at Girimint Colliery. Although coal bumps are not a major hazard at present in India, coal mining is bound to go deeper, increasing the hazard in future.

DETAILS OF THE CASE

At this mine, Dishergarh seam is 2.4–2.7 m thick, dips at 1 in 5, has a depth of 140–360 m and is being worked by bord and pillar system. Another seam Manoharbahal, is 180 m above and is waterlogged. There is no lower workable seam. Panel H-1, shown in Fig. 1, was being depillared with caving when on 18/8/84 severe pillar spalling occurred, killing one worker. The spalling was associated with a big sound which was reportedly heard on the surface. The depth of cover in this panel is + 300 m.

Underground inspection revealed the following facts:

132

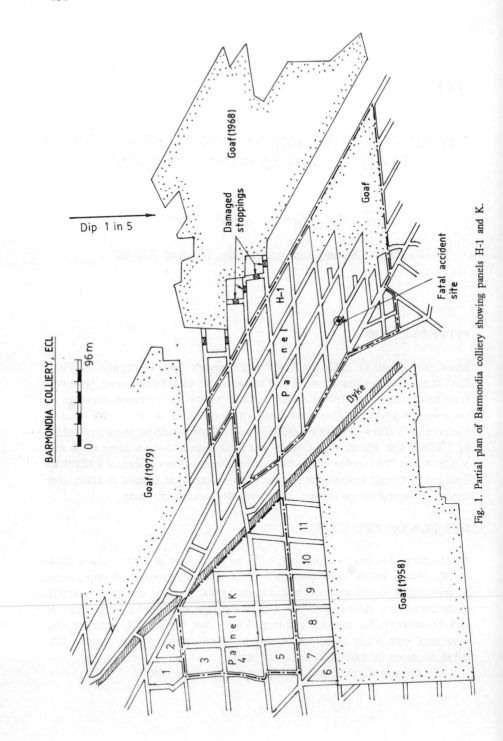

Fig. 1. Partial plan of Barmondia colliery showing panels H-1 and K.

1) Pillars on the rise side had spalled more than dip-side pillars.

2) Explosion-proof stoppings between the rise-side goaf and Panel H-1 had been damaged.

3) The immediate roof shale up to 0.75 m had fallen more on the rise side.

4) Roadways at places had widened to 7 m due to spalling.

5) Since the spalling was severe, it was difficult to ascertain whether coal pieces had flown over long distances, as in a bump.

6) Except for part of the immediate shale, the roof in the goaf of Panel H-1 had not fallen.

Points (1) and (2) indicated that high rock stresses had developed at the rise side. The only probable cause was a major roof fall in the goaf standing (since 1968) to the rise side of Panel H-1 (Fig. 1). This goaf had a total area of about 30,000 m². Such large areas cannot stand hanging even under the hardest coal measures [1], unless they are aided by pillar remnants and panel shape. Considering the practise of leaving 30% coal as pillar remnants in depillaring at this mine, this could be considered probable, especially since the strata are hard. This was thus classified as a "shock" bump, rather than a "pressure" bump, created by the dynamic action of roof fall in the neighbouring goaf.

The problems before the mine management were: (a) How should Panel H-1 be extracted further? (b) What would be the safe manner of extracting the neighbouring panel K (Fig. 1)? The conditions associated with these panels were:

Panel H-1

1) Considerable spalling had already taken place and the pillars might spall further due to goaf abutment pressure.

2) The collapse in the rise-side goaf, which was earlier supported by pillar remnants, might have caused further increase in rock pressure.

3) Rhombic pillars.

Panel K

1) Since the panel is large, the waterlogged seam 180 m above might be affected.

2) A large goaf dipwards of this panel which, if the roof collapsed, could cause a similar incident as in Panel H-1.

STUDIES FOR BUMP PRONENESS

A seam is considered bump prone if it satisfies the following well-known natural or uncontrollable conditions simultaneously [2, 3].

a) energy index of coal, $W_{ET} \geq 2$;

b) strong roof and floor; and

c) depth $\geq$ 300 m.

134

To these may be added the mining or controllable factors, viz., a massive roof fall in the same or adjoining panel, creation of high abutment stresses in the working area etc.

The depth of these panels is + 300 m and it is seen that condition (b) is also satisfied, judging by the RQD plot of the roof to a height of 16 m (Fig. 2). The roof has 75% RQD on average, which comes under the category of "cavable with difficulty".

It was then decided to study condition (a) for bump proneness.

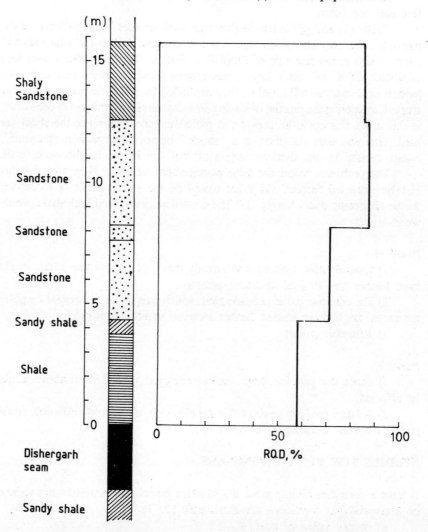

Fig. 2. Strata borehole section for Dishergarh seam and RQD values.

Coal Behaviour

The energy index (W_{ET}) is defined in terms of the recoverable strain energy of a coal, i.e., the ratio of recoverable and irrecoverable energies, as obtained from the stress-strain characteristic. The definition appears quite logical but may not be sufficiently comprehensive.

It has also been suggested [4] that the violence of a burst may be related to the post-failure slope of the rock. Earlier, the Czechs came up with the elastic limit of the stress-strain curve as the governing factor, supplemented by factors of roof strength and bed thickness [5]. According to them, for bump proneness,

$$L_e = \frac{\varepsilon_e}{\varepsilon_t} \geq 0.7$$

where ε_e and ε_t are elastic and total strains respectively and L_e is the ratio of the two.

Figures 3 and 4 show the test results for energy index and elastic limit.

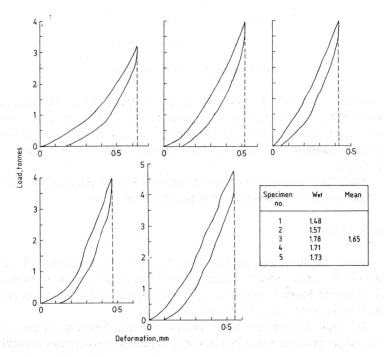

Fig. 3. Loading-unloading curves and energy index values for Dishergarh coal.

136

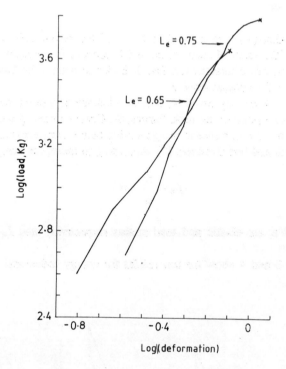

Fig. 4. Elastic limit of Dishergarh coal.

The energy index was obtained on average as W_{ET} = 1.65. The elastic limit was also obtained, the average value being L_e = 0.7. These tests thus indicate that the seam may be marginally bump prone. It should be mentioned that for these tests samples were obtained from various horizons of the seam to get an average.

Since the other two conditions of hard strata and depth are also prevalent, the seam could be classified as marginally bump prone.

Mining Factors

Cook *et al.* [6] have suggested in 1966 that the energy release rate *(ERR)*, defined as the strain energy released per unit area mined, should be minimised by mine layout design to minimise rockburst hazards. This subsequently proved correct in the gold mines of S. Africa.

By analysing stresses around tabular deposits, Salamon [7] has shown that the major principal stress (σ_1) is related in *ERR* in a plane strain situation by

$$\sigma_1^2 = \frac{2E}{(1-v^2)h} \; ERR \tag{1}$$

where E, v are elastic constants of the reef and h is the reef thickness. This equation shows that a reduction in stress or ERR will have an identical effect on rockbursts. A reduction in the reef thickness or working height will also reduce ERR and hence rockbursts. A decrease in the extraction width will also act in the same manner because it will induce smaller values of σ_1. Partial extraction, using smaller stope widths and intervening pillars, has proved to be very popular in S. Africa. Since coal seams are also tabular deposits the same logic applies.

STRATEGIES FOR EXTRACTION

Panel H-1

Since the pillars were rhombic and already reduced due to spalling in this panel, it was decided to reduce extraction width to decrease ERR. Also, it was decided to avoid a main fall to prevent a shock burst, the pillars being already marginally stable. The width which would cause a main fall was obtained for RQD = 75 % as [1]

$W = 0.59 \; RQD + 5.2$ m
$\quad = 49$ m

If two rows of pillars were extracted, the width W would be 60 m. Hence the panel was extracted by taking every alternate row, thus keeping extraction width at 32 m. Since the pillars were already reduced, their sides were supported by "stitching" with grouted wire ropes and wooden lagging to tighten the rope (Fig. 5). Such stitches were placed vertically from roof to floor, the ropes being grouted 2.5–3 m deep. The horizontal distance between stitches was 1.5–2 m and they were placed only on the sides facing extraction.

Extraction of Panel H-1 was completed successfully with no incident of spalling or bump, nor did the roof cave.

Panel K

The pillars in panel K were square and larger than in panel H-1 although a large goaf has existed since 1958 to the dip side. The condition of this goaf was unknown and so it was decided to take the precaution of side stitching of pillars, in case the old goaf collapsed.

Considering the waterlogged workings in Manoharbahal seam 180 m above, the equivalent rectangular area of the panel was calculated. This rectangle was then converted to an infinitely long rectangle (plane-strain situation) using the

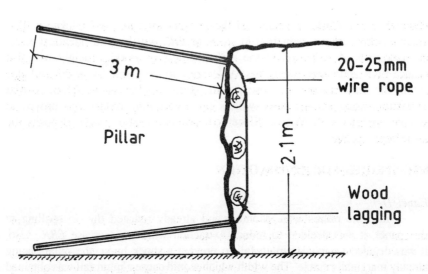

Fig. 5. Rope stitching of pillar sides.

theory of plates. The width of this rectangle was estimated as 90 m, which gave the panel width-parting rock ratio between the two seams as 90/180 = 0.5. Subsidence observations in the same seam elsewhere in Raniganj Coalfield [8] indicated that this ratio comes under the non-effective width of extraction, which would cause no subsidence.

ERR was kept to a minimum by reducing the working height to 2.1 m (leaving roof coal) and also by leaving 30% coal as pillar remnants in the goaf. There was a second reason for leaving roof-coal to control the behaviour of the thin shale bed. Leaving coal ribs was also suggested to prevent roof caving, to obviate the occurrence of shock bumps.

In order to estimate what would be the condition of pillars under abutment action in panel K, rock pressure distribution over the pillars was obtained using the CMRS finite difference computer programme ROCP [9]. The average pressure over each of the boundary pillars numbered 1-11 was also computed. The compressive strength of the seam was found from lumps collected from various places and cut into 2.5 cm cubes. The average strength was obtained as 36 MPa. The following CMRS formula for strength of coal pillars [10] was used with the computed rock pressures for obtaining safety factor values (Table 1):

$$S = 0.27\sigma_c h^{-0.36} + \frac{H}{160}\left(\frac{w}{h} - 1\right) \text{ MPa}$$

where S = pillar strength (MPa)

σ_c = compressive strength (MPa)

h = working height (2.1 m)

H = depth of cover (300 m)

w = pillar width.

The safety factor of the pillars was found to be reasonable enough. Hence it was not necessary to adopt the limited-span method as in panel H-1.

Table 1. Pillar safety factors around panel K

Pillar No.	Width (m)	Strength (MPa)	Load (MPa)	Safety factor
1	32	34.14	11.1	3.07
2	18	21.64	15.5	1.39
3	34	35.92	15.2	2.36
4	36	37.71	16.1	2.34
5	34	35.92	15.0	2.39
6	18	21.64	15.0	1.44
7	28	30.56	15.1	2.02
8	35	36.82	18.6	1.98
9	37	38.60	20.8	1.85
10	39	40.38	20.5	1.97
11	40	41.28	18.8	2.19

Drilling Yield Tests

As a precautionary measure, drill yield tests were conducted to locate a dangerous bump situation, if any. A 42 mm diameter bit, readily available for coal mines, was used to drill holes in the pillar sides within one pillar distance (about 30 m) of the edge, which was the expected high-stress zone. Gummings were collected every metre and the volume measured. If the volume is greater than 6 litres/m a dangerous situation exists as per German norms [11] and the distance at which this occurs is to be classified according to Polish norm [12] as:

> 3.4 h   not dangerous

3.4–1.5 h   dangerous

< 1.5 h   critical

A typical plot of a 6.75 m hole in Fig. 6 shows no danger due to bumps. The yield of gummings was generally observed to be less than 3 litres/m.

Thus Panel K was also extracted successfully.

GENERAL DISCUSSION

A case of perhaps only marginal bump proneness has been presented. While

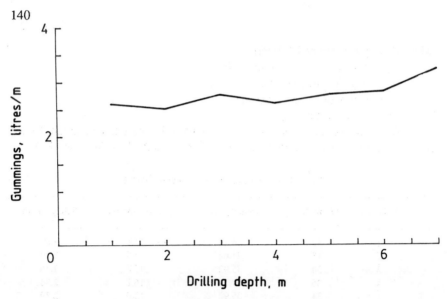

Fig. 6. Drill yield test result of a typical hole in panel K.

taking up the investigations at this mine, the authors had the feeling that bump proneness has not been very satisfactorily defined to date. The conditions of hard coal, strong roof and floor and depths of + 300 m are at best qualitative. For example, what is meant by a strong (hard) roof? The Czechs have attempted to quantify it in terms of bed thickness and strength. In coal mining, roof hardness is clearly synonymous with cavability and a classification index such as RQD, which has shown a good correlation with cavability [1], might, perhaps, be more suitable. Another useful parameter could be the caving index which combines RQD and compressive strength [13]. But the question still remains as to what is that limit of RQD which proves dangerous over a bump prone seam. The authors suggest a tentative limit of 60%.

Similarly, it is often said that a soft bed (for example, shale) in immediate contact with the seam, above or below, alleviates the bump situation, even if the rest of the strata are hard. What should be the minimum thickness of such a bed for such alleviation?

Coming to coal behaviour, the energy index, elastic limit and post-failure stiffness have been put forward as measures of bump proneness. Besides these three, two more factors might have to be considered. It is well established by now that if the abutment pressure is thrown further into the solid (by controlled blasting, auger drilling, water infusion or because the coal is soft in the first place), a bump hazard reduces or does not exist. This means that if we have a large failed coal zone in the abutments, bumps may not occur. This brings us to the equation given by Wilson for failed coal zones between a rigid roof and floor [15].

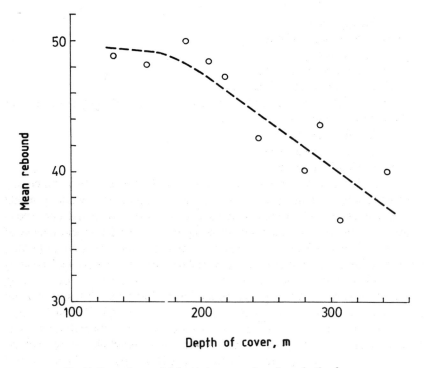

Fig. 7. Dependence of Schmidt hammer rebound on depth of cover.

$$x = \frac{h}{F} \ln \frac{\gamma H (q-1)}{\sigma_c}$$

where x = width of failed coal zone

$$F = \frac{q-1}{\sqrt{q}} + \frac{(q-1)^2}{q} \tan^{-1} \sqrt{q}$$

$q = (\sqrt{1 + \mu^2} + \mu)^2$

$\mu = \tan \phi$ = coefficient of internal friction

γ = unit rock pressure

This equation is given when there are no supports against the coal side. The elasto-plastic theory, of which this equation is one solution, shows that the extent of rock failure depends on two material properties, viz, compressive strength (σ_c) and angle of internal friction (ϕ). These two material properties should, therefore, influence bump-proneness.

We can thus summarise the factors which contribute to a bump-prone situation in a coal seam:

1) Energy index.

2) Post-failure stiffness.

3) Elastic limit.

4) Compressive strength.

5) Angle of internal friction.

6) Roof cavability defined by a suitable index, e.g., RQD (the same index being applied to the floor).

7) Depth of cover.

8) Energy release rate.

9) Mining-induced stress.

The first seven are inherent factors, while the last two which are complementary, are mining factors dependent on mine layout, working height and goaf treatment. Of the inherent factors, the first five should correspond to the seam *in situ*, i.e., they should be specified considering the degree of jointing. Barton's rock quality Q [15] or Bieniawski's rock mass rating (RMR) [16] can define jointing quite satisfactorily, but the question of correlating laboratory determinations with the *in situ* values still remains. Another way of obtaining the *in situ* values could be by underground tests. The Schmidt hammer might prove quite useful in this respect, e.g., for *in situ* strength [17] and energy index [18]. It should, however, be used with care as the rebound values may be considerably biased by the depth of cover, as reported elsewhere [10]. At Barmondia, for example, Schmidt hammer type N was used to obtain rebound data. Vertical channels 1.5 m deep and 1 m wide were prepared in pillar sides. Grids 30 cm wide, with 7.5 cm spacing of grid points, were drawn on two walls of the trenches at right angles. Figure 7 shows the variation of mean rebound against depth of cover. A noticeable fall in rebound value occurs because the failure zone increases with cover while the depth of testing channel remains constant. This indicates therefore that increasingly crushed coal is tested as depth increases. Under shallow covers, we had a 'nominal'' failure zone, which occurred because of blasting, and which was eliminated in this instance by a 1.5 m deep channel.

Much research needs to be done for defining bump proneness of coal seams. The nine factors delineated in this section may point out areas of further work. CMRS intends to take up a major research project on this subject.

ACKNOWLEDGEMENTS

The authors are grateful to Dr. M.N. Das, Mr. D. Barat, Mr. R.K. Prasad and Mr. R. Rao for carrying out the various investigations at Barmondia. Special thanks are also due to the Barmondia management, particularly Mr. B.B. Singh, Dy. CME. The opinions expressed are those of the authors and not necessarily of CMRS.

REFERENCES

1. Sheorey, P.R. 1984. Use of rock classification to estimate roof caving span in oblong workings, *Int. J. Min. Eng.*, vol. 2, pp. 133-140.
2. Szecowska, Z., J. Domzat and P. Ozana. 1973. Energy index of coal liable to bump, *Prace GIG*, kom nr. 594.
3. Neyman, B.Z., Z. Szecowska and W. Zuberek. 1972. Effective methods for fighting rockbursts in Polish collieries. *5th Int. Strata Control Conf., London,* 1972, Paper 23.
4. Farmer I.W. 1985. *Coal Mine Structures.* Champman and Hall, London.
5. Zamarski, B. and J. Franek. 1977. Theoretical principles, organisation and mining practise for control of ground in subsurface mines in Czechoslovakia. *6th Int. Strata Control Conf., Banff.* 1977.
6. Cook, N.G.W., E. Hoek, J.P.G. Pretorius, W. D. Ortlepp and M.D.G. Salamon. 1966. Rock mechanics applied to the study of rockbursts, *J.S. Afr. Inst. Min. Metall.*, vol. 66, pp. 435-528.
7. Salamon, M.D.G. 1983. Rockburst hazard and the fight for its alleviation in S. African gold mines. *Symp. Rockbursts: Prediction and Control,* London, pp. 11-36.
8. Saxena, N.C. and B. Singh. 1986. A status report on problems of subsidence in mining areas, *J. Mines, Metals, Fuels,* pp. 435-445.
9. Sheorey, P.R., M.N. Das and B. Singh. A numerical procedure for rock pressure problems in level seams. *Symp. Strata Mechanics, Newcastle-upon-Tyne,* pp. 254-259.
10. Sheorey, P.R., M.N. Das, D. Barat, R.K. Prasad and B. Singh. Coal pillar strength estimation from failed and stable cases. *Int. J. Rock Mech. Min. Sci. Geomech.* Abstr. (accepted, 1987).
11. Kastenbahn. Personal communication.
12. Peng, S.S. 1979. *Coal Mine Ground Control.* John Wiley, New York.
13. Sarkar, S.K. and B. Singh. 1985. *Longwall Mining in India.* Published by Mrs. S. Sarkar, Dhanbad.
14. Wilson, A.H. 1980. Pillar stability in longwall mining, pp. 85-95. In: *Ground Control in Longwall Mining and Mining Subsidence.* SME-AIME, New York.
15. Barton, N., R. Lien and J. Lunde. 1974. Engineering classification of rock masses for the design of tunnel support, *Rock Mech.*, vol. 6, pp. 189-236.
16. Bieniawski, Z.T. 1976. Rock mass classifications for rock engineering. *Symp. Exploration for Rock Engg., Johannesburg.*, vol. 1, pp. 97-106.
17. Sheorey, P.R., D. Barat, M.N. Das, K.P. Mukherjee and B. Singh. 1984. Schmidt hammer rebound data for estimation of large-scale *in situ* coal strength. *Int. J. Rock. Mech. Min. Sci. Geomech.*, vol. 21, pp. 39-42 (Abstr.).
18. Kidyibinski, A. 1981. Bursting liability indices of coal. *Int. J. Rock Mech. Min. Sci. Geomech.*, vol. 18, pp. 295-304 (Abstr.).

11

MOST PROBABLE MECHANISM OF ROOF ROCKBURSTS IN COAL MINES OF BYTOM SYNCLINE

Antoni Goszcz

Central Mining Institute, Katowice, Poland

INTRODUCTION

Inadequate understanding of the mechanism of rockburst is one of the main reasons for unsatisfactory progress in combating this hazard. In the literature this mechanism is defined in various ways as, for example, the violent crushing of a seam remnant, rock failure in the rock mass, crack propagation exceeding the critical state, etc. Without disputing these postulations it should be noted that this divergence of views has had consequences in the form of differing, and often controversial opinions, on methods for prevention and prediction of this hazard. One may mention here the ongoing discussions on the effects of extraction rate on the state of hazard, advisability of weakening the sidewalls of workings by blasting or water infusion, inadequacy of prediction methods, etc.

This divergence has its own historical background. Most of the opinions derive from the period when roof rockbursts were only of marginal significance. Today, the situation is reversed and roof rockbursts are a major hazard while seam rockbursts occur only rarely. This is the main reason for the limited effectiveness of forecasts and preventive measures, which in the past have been virtually oriented exclusively towards seam rockbursts.

In the 1980s, extensive studies were carried out on the mechanism of mining tremors and rockbursts in the Bytom Syncline (Marcak, 1985). It was ascertained that the mechanism of high-energy tremors and rockbursts can be described by the stick-slip geodynamic model, used in explaining the earthquake mechanism (Scholtz, Molnar, Johanson, 1972; Byerlee, Summers, 1975; Nur, 1978). Obviously, the scale of rockbursts phenomena in mines and the process itself differs substantially from that of earthquakes. Nevertheless, certain mining phenomena, especially roof rockbursts, can be explained by the stick-slip model.

ELEMENTS OF GEOLOGICAL STRUCTURE OF BYTOM SYNCLINE, UPPER SILESIAN COAL BASIN

The Bytom Syncline is an extensive brachsyncline of ellipsoid shape with the longer axis lying roughly E-W (Fig. 1). The Carboniferous formations are covered by comparatively regularly formed Quaternary and Lower Trias formations up

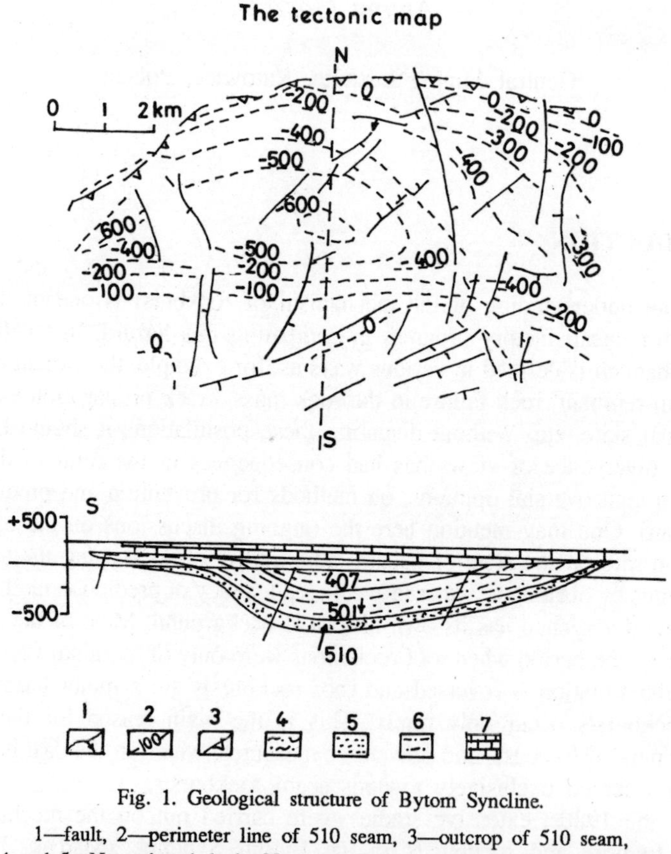

The tectonic map

Fig. 1. Geological structure of Bytom Syncline.

1—fault, 2—perimeter line of 510 seam, 3—outcrop of 510 seam,
4 and 5—Namurian A beds, Namurian B beds (so-called "saddle beds"),
6—Namurian C beds, 7—triasic beds.

to 180 m in thickness. The deeper Carboniferous strata are folded to form a syncline. The northern flank is inclined at an angle of up to 20°. The southern flank is steeper with an angle of dip even up to 45°. From the lithological point of view, the Carboniferous formations of the Bytom Syncline were formed in a manner typical for the Upper Silesian Coal Basin, creating a stack of sandstone-mudstone-shale layers with interleaved coal seams that are exploited in several mines. Extracted here are Namurian seams (Namur B and Namur C). Mining is conducted at depths varying from 400 to 900 m, usually in longwall systems with caving. Hydraulic fill is used only rarely, when necessary for surface protection. Thick seams, i.e. those over 4 m, are extracted in slices parallel to the floor. The chief direction of advance of the extraction front is due E-W or W-E.

MINING TREMORS AND ROCKBURSTS IN THE BYTOM SYNCLINE

Mining tremors and rockbursts occur very frequently during the extraction of coal seams in the Bytom Syncline. These rockbursts are principally roof rockbursts caused by deformations of thick beds of strong sandstones occurring in the main roof over the extracted seams.

The annual average for tremors recorded by the mining seismologic stations in this area is:

Tremor energy (J)	No. of tremors
$10^5 \div 10^6$	670
$10^6 \div 10^7$	100
$10^7 \div 10^8$	12
over 10^8	2

There are proportionally more tremors with energies below 10^5 J. High-energy mining tremors can cause rockbursts. During recent years, rockbursts have occurred several times per year. High-energy tremors can also cause damage to surface structures since mining is conducted under the city of Bytom and industrial plants.

The mines are equipped with their own seismologic apparatus enabling them to locate the foci of tremors (or at least to determine the co-ordinates of the epicentre) and to estimate the quantity of energy released. The equipment in use permits an accuracy of ± 30 metres in calculating the co-ordinates of the epicentre.

ADAPTATION OF THE STICK-SLIP MODEL FOR DESCRIPTION OF ROOF ROCKBURSTS

According to the stick-slip model, rockbursts can occur only when the rock mass overlying the seam being mined (or already mined out) is divided into blocks and, at the same time, the dividing surfaces are oriented not only transversely but also parallel to the bedding. The existence of dividing surfaces transverse to bedding is obvious if the surfaces of discontinuity (faults, jointing, cleavage, etc.) and the fractures formed above the extracted seam due to mining operations are taken into account (Fig. 2). The latter are specially well developed in the edge areas where work ceased sometime ago. Therefore, zones in the vicinity of remnants, abutment pillars, gobs, edges etc., have a high rockburst hazard.

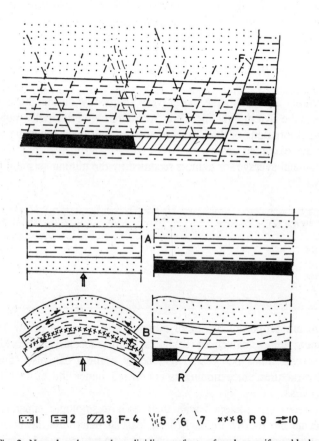

Fig. 2. Natural and secondary dividing surfaces of rock massif on blocks.

1—strong rock, 2—weak rock, 3—excavation 4—fault, 5—post-mining fractures, 6—cleavage, 7—joint, 8—mylonitised zone, 9—horizontal separation, 10—decollement.

Dividing surfaces parallel to bedding can be the result of various geological and mining causes. Geological causes most frequently include folding of several strata of different tractable properties. In this situation, at the surface between tractable and non-tractable strata (e.g., where a bed of competent sandstone and a layer of mudstone are in contact, a rupture (decollement) occurs with a mirror-finished slip surface. The slip surface may also appear as a thin (of the order of 20 cm) highly mylonitised layer, colloquially called "mydlik," which forms in the layer of lowest shear strength.

In all cases (decollements and slip and mylonitisation), dividing surfaces parallel to bedding occur due to the existence of shear stresses associated with transverse bedding of layers as a result of folding during orogenesis. Dividing surfaces in the rock body parallel to bedding may also occur as a result of mining operations. The simplest form of division is separation under their own weight of strata overlying the mined-out areas when after extraction of a seam a thick layer is left without support (Fig. 2). A surface of discontinuity parallel to bedding may also appear as a result of shearing due to interlayer movements ahead of the face line, especially where rock with inclusions of plastic mudstone occurs in the roof. It should be underlined that the occurrence of surfaces of discontinuity parallel to bedding is most likely where the massif is composed of thick rock layers of highly differentiated strength (competent and weak rock). This is why thick beds of competent sandstone or mudstone overlying the extracted seam represent a very serious risk of rockburst occurrence. At the contact zone of such layers with underlying more tractable shales, separation of strata in the rock body is highly probable.

The processes described above cause the division of the massif into blocks and with sufficiently competent rock occurring in the roof, in some mining situations such as under- or overpassing the edge, approaching the gob, extracting remnants, approaching faults, etc., the hazard of roof rockburst is particularly significant. The mechanism of such rockbursts in a typical mining situation when approaching the gob is outlined in Fig. 3.

Extraction of coal under block B (Fig. 3—right) up to edge K will create steep fractures above this edge. Rocks in block B, overlying the extracted seam and forming the immediate roof, are lowered by a depth (w), where w = m. a, a being the subsidence factor. If, however the main roof is formed of competent rocks, it will be difficult for these rocks to subside due to "wedging" in the cracks. Then a horizontal separation (R) occurs at the contact surfaces between weak and competent rocks, the rocks are "suspended" and the whole system is highly unstable. This unstable equilibrium could easily be disturbed due to:

— Prolonged load effect. The long-term strength of rock is significantly lower than the ultimate strength.
— Reduction in friction force (T) in the crack as a result of reduction of friction coefficient.
— Reduction in friction force (T) due to the appearance of tensile stresses in the crack zone caused by mining operations adjacent to the edge.

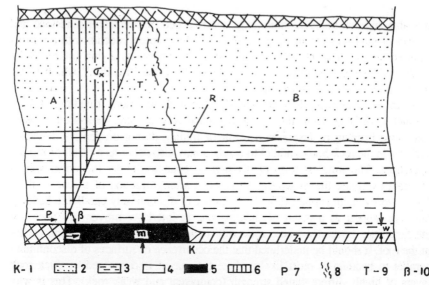

Fig. 3. Mechanism of a roof rockburst.

1—edge, 2—main roof, 3—nether roof, 4—works, 5—coal seam, 6—zone of tension,
7—working face, 8—fracture zone, 9—force of friction on
the crack, 10—angle of influence.

In the first two cases, the loss of stability in the crack will appear as a typical tremor over the gob while in the third case, as a typical mining rockburst in the vicinity of the mined-out area.

The mechanism of such a rockburst can easily be explained if it is remembered that ahead of the working face (F) tension (+ σ_x) occurs. The zone of tension is bound by a line inclined at an angle (β) (Fig. 3) to the horizontal plane. If the face (F) is so close to crack no. 8 that part of it actually lies in competent rock in a tensile zone, then the highest drop in friction force (T) occurs in the crack. This may cause unlocking of the crack and sliding of block B over a distance not greater than Δz. A slip of this kind releases the following quantity of potential energy:

$$E_p \leq \frac{1}{2} \Delta z . m . g$$

This may be high enough to trigger a rockburst. For example, if a 50 m × 50 m × 50 m³ block suffers a slip of 5 cm (Δz = 10 cm), the quantity of energy released will be sufficient to cause a tremor of energy in the range $10^6 \div 10^7$ J.

Assuming this described mechanism, the occurrence of a mining tremor is possible only when there is separation (R) and also "suspension" of the rock block. Hence, the occurrence of two rockbursts at the same place is also possible only when the first one has not eliminated the separation or when the separation has become "rejuvenated" by subsequent mining operations. This mechanism of roof rockbursts should be considered a hypothesis; however, analysis of rockbursts actually occurring in coal mines appears to confirm its validity.

SOME PROBLEMS OF ROCKBURSTS PREDICTION

According to the stick-slip mechanism, roof rockbursts may be considered uncontrollable sliding of a rock block along the steeply inclined surface of a stratum discontinuity. It is preceded by:

— initial sliding up to the moment of locking of this movement and "suspending" of the block;
— preparation of the surface for slip by breaking off of the weakest dents (Fig. 3) and smoothening of asperities; and
— appearance of local dilatancy in a comparatively narrow zone, close to the slip surface.

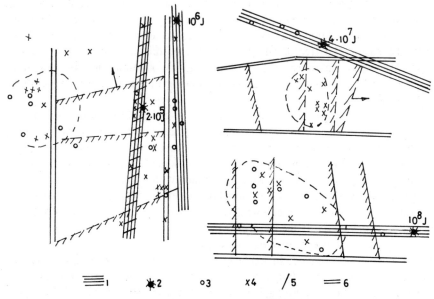

Fig. 4. Seismological history of rockbursts.

1—sliding zone, 2—rockburst foci, 3—tremors of E = 10^5–10^6 J,
4—tremors of E = 10^4–10^5 J,
5—position of working faces, 6—galleries.

Sometimes an increase in dampness and rock "sweating" may occur in the hazard zone.

Having considered the type and character of phenomena preceding a roof rockburst, it is obvious that mining seismology can only provide useful result if tremor foci locations are analysed in detail. Smoothening of each asperity and breaking off of projections could prepare a slip surface and could also cause a mining tremor of lower energy (a kind of foreshock). Clearly the shock sources are concentrated in a narrow zone of the future slip. This has been confirmed by study of the seismological history of rockbursts in mines of Bytom Syncline (Fig. 4).

Two categories of tremors can be distinguished here:

1) Randomly distributed tremors forming a "cloud".

2) Foreshocks whose sources are clustered within a narrow zone where the focus of the main tremor is also found.

The fact that rockburst foreshock sources are distributed along a narrow zone raises hopes for successful prediction of a hazard state based on mining seismological data. Other methods, e.g., drilling of small diameter boreholes and seismo-acoustic methods are currently of doubtful reliability. If this hypothesis for the mechanism of roof rockbursts proves true, major changes need to be introduced in the currently applied preventive measures. This issue, however, requires a separate analysis. Therefore, it is only possible to postulate guidelines in certain particularly important areas. First is the problem of choice of mining system. A roof rockburst can occur only when separation of strata takes place over the seam. The tremor energy will then be directly proportional to Δz (Fig. 3). Hence the safest system of mining in conditions of roof rockburst hazard will be one using incompressible stowing in which case no separation of roof strata takes place.

The destressing method, i.e., mining out a so-called destressing seam, should have a useful effect on the roof rockburst hazard, but on the contrary it may have a detrimental effect, causing division of the rock body into blocks. Extraction of the upper seam (overmining) creates an additional dividing surface parallel to the bedding (old works in Fig. 3), bounding the block from the top and reducing cohesion in the massif. It should be remembered however, that extraction of a destressing seam ensures effective destressing of coal seams and reduces liability to rockbursts. If the described mechanism of roof rockbursts is sound, weakening of coal sidewalls in narrow workings by means of blasting and water infusion should be limited. In roadway workings, the weaker the sidewalls the greater the effect of rockbursts, and vice versa. Therefore, if the sidewalls of workings are sufficiently strong, lower energy rockbursts need not have disastrous effects. If, however, these sidewalls are subjected to "blasting", even a small shock could be disastrous in the blasted or infused sidewalls.

REFERENCES

Byerlee, J.D. and R. Summers. 1975. Stable sliding preceding stick-slip of fault surface in granite at high pressure. *Pageoph.* vol. 113.

Marcak, H. 1985. Geofizyczne modele rozwoju niszczenia gorotworu poprzedzajacego tapniecia i wstrzasy w kopalniach. Publ. Inst. Pol. Acad. Sc. M-6, 176.

Nur, A. 1978. Nonuniform friction on a physical basis for earthquake mechanism, *Pageoph.* vol. 116.

Scholtz, C., P. Molnar and T. Johanson. 1972. Detailed studies of frictional sliding of granite and implications for earthquake mechanism. *Journ. Geoph. Res.,* vol. 77.

12

SAFE AND EFFECTIVE TECHNOLOGY FOR MINING OF BUMP-PRONE DISHERGARH SEAM AT CHINAKURI MINE

T.N. Singh and B. Singh

Central Mining Research Station, Dhanbad

INTRODUCTION

The phenomenon of bump signifies rock and coal displacement with sharp noise and some violence. In some cases, when the geological disturbances were in close vicinity to the exposed surface, violence increased causing detachment of roof and seam and damaging supports as in a rockburst. The bumps as such cover all phenomena associated with instantaneous failure of Coal Measure formation due to release of stresses. The Dishergarh seam occupies the top position among seams prone to bump in Indian mines. This seam has been extracted by bord and pillar and/or longwall in conjunction with stowing when the workings were associated with occurrences of bumps of different intensity. The seam has been worked in slices at Chinakuri mine where its thickness exceeded 4.5 m, while only the bottom section (2 m thick) was worked in Pits 1 & 2 sterilising a huge amount of coal in the top section. Equivalent material model studies were undertaken in the laboratory during 1975 to 1980 to study methods for extraction of full seam independently as well as in collaboration with the PSWETMETPROMEXPORT team at CMRS, Dhanbad. A proposal of composite slicing was ultimately formulated to facilitate extraction of 4.8 m thick Dishergarh seam below 700 m cover in the eastern sector. This paper deals with different aspects of the methods proposed for full-seam working in Chinakuri Pits 1 & 2.

BUMP IN INDIAN MINES

The phenomenon of bump in some Indian coal mines is known from very early days. Nearly 94 occurrences of bumps were reported during 1944 to 1964 in different coal mines causing 135 fatal accidents [1]. The depillaring operation accounted for 85% of the total accidents including 58.4% on extraction faces and 26.6% in split galleries. The longwall faces of Parbelia Colliery accounted for 9.5% of the total bumps. The occurrence of bumps increased with increase in depth (Fig. 1). The deepest workings of Parbelia mine had a number of bumps even with the longwall method. The other deep mine, Dhemomain, working by conventional bord and pillar system met the highest number of bumps at 480 m depth cover. This was followed by Bejdih Colliery. The Koithi seam working in Sripur mine at depth cover of nearly 300 m had also shown considerable proneness to bump. In Dhemomain alone, bumps caused 6 fatal accidents in 1944, 4 in 1945 and 12 in 1952 due to deterioration of pillars and presence of hard sandstone over the seam. The pillars deteriorated because of spontaneous heating, flooding and dewatering. In other mines, bumps were mainly associated with the process of stooking or delay in the process of stowing. The Karharbari seam of Giridih mine had a number of bumps though the workings were comparatively shallow.

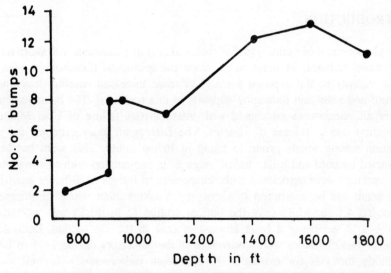

Fig. 1. Occurrence of bumps with increase in depth.

Dishergarh seam at Chinakuri Pits 1 & 2 is being worked below 700 m depth cover, the thickness of seam varying from 3 m to 4.9 m. The coal seam is overlain by 17 to 31 m thick sandstone (Fig. 2) of compressive strength varying from 500 to 750 kg/cm² (Table 1). The seam has been reported to be highly bump prone according to Indian standard and marginally bumpy according to the standard of Soviet mines. In the light of difficult massive nature of strata and poor remedial measures, the seam was taken as highly bump prone. It was, therefore, not advisable to extract the seam by bord and pillar system or even by conventional longwall to the full height without supports of high capacity. Polish observations [2] confirmed 50% bumps on faces while working remnant pillars. Investigations undertaken by the United States Bureau of Mines revealed nearly all bumps occurred in the area of pillar extraction. Of 117 bumps, nearly 66.7% were at the pillar line and 12% in the depillaring zone. Thus nearly 80% of the bumps were in the pillar extraction zone. Occurrences of bumps during different operations of pillar extraction are given in Table 2.

Table 1. Physico-mechanical properties of core samples drilled at 47 East dip 11-1/2 level at Chinakuri Pits 1 & 2

Formation	Distance from Dishergarh	Mean compressive strength (σ_c), kg/cm²	Mean shear strength (τ) kg/cm²	Mean tensile strength parallel to bedding plane, kg/cm²	Mean tensile strength perpendicular to bedding plane, kg/cm²	Bulk density
Shale	0–0.55	693	63	9	65	2.55
Intercalation of shale and sandstone	0.55– 2.32	690	131	11	56	2.36
Sandstone (fg)	2.32– 7.60	738	167	14		2.60
Sandstone (fg)	2.32– 7.60	860	124	12		2.58
Sandstone (fg)	2.32– 7.60	763	141	18	72	2.58
Sandstone (fg) white	7.60–12.66	957	112	31	72	2.72
Sandstone (mg) grey	12.66–17.68	823	129	40	70	2.51
Sandstone (mg) grey	12.66–17.68	674	111	18	67	2.56
Sandstone (fg) grey	17.66–18.74	501	116	23	57	2.46

Longwall system of mining has been found to be comparatively safe in bump-prone seams. Polish observations confirm safe extraction of a bump-prone seam by longwall, preferably by advancing in conjunction with stowing. The height of extraction in most cases were preferably as minimal as possible in order to reduce occurrences as well as intensity of bumps. The relation of working height to number and intensity of bumps in Soviet mines [3] (Fig. 3) revealed an abrupt increase in bumps at working heights over 1.1 m.

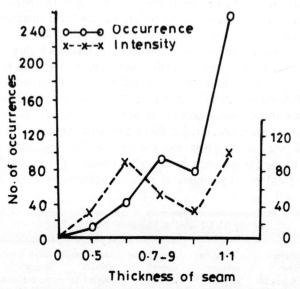

Fig. 2. Coal measure formation in and around Chinakuri Pits 1 & 2.

Table 2. Analysis of incidence of bumps in coal mines in the United States

Mining operation	No. of bumps	Per cent	Bumps which occurred on pillar line points listed in column (2)	
			Number	Per cent
Open-end pillar extraction	23	19.7	14	60.8
Stabling pillars in abutment area	31	26.5	23	74.2
Developing in abutment area	14	12.0	8	57.00
Splitting pillars in abutment area	9	7.7	7	77.8
Pillar being mined, bump not at face	2	1.7	1	50.00
Pillars back from extraction line not being mined	11	9.4	6	54.5
District bump	6	5.1	6	100.0
Pillar and stall pillar recovery	5	4.3	3	60.0
Recovery of pillars surrounded by goaf and small pillars partly mined	3	2.5	3	100.0
Miscellaneous	4	3.4	2	50.0
Insufficient information	9	7.7	6	66.7
	117	100.0	79	

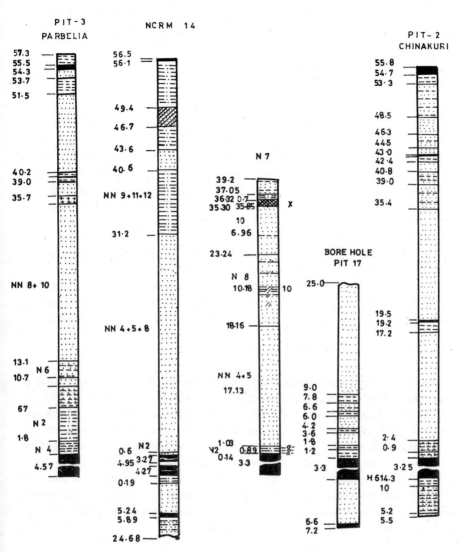

Fig. 3. Occurrence of bumps with respect to working height.

MINING PROPOSITION

Exploitation of a seam prone to bump, gas and rockburst at a critical depth might lead to dynamic strata failure and outburst of coal. Such phenomena were observed in Baberia, Kizelova and Pshibram mining regions of the FRG, USSR and Czechoslovakia due to strong and elastic coal seams, thick strong immediate roof, great mining depth and discontinuity-like folds. Operational parameters such as presence of pillars in adjacent workings and mining of two faces approaching each other were invariably responsible for such problems. Experiences in Indian mines in no way differ from reported observations. USBM observations have confirmed 2 to 4 times more bumps in bord and pillar working compared to longwall faces. In Poland, longwall faces account for only 10.5% of the total bump occurrences. It is normal convention that where the thickness of cover exceeds 400 m and the seam is prone to bump, longwall system of mining with or without stowing should be preferred. The French Commission on Bumps [4], however, opined that longwall working with caving was unsuitable for seams over 4 m thick. Slow and regular movement of straight face, effective packing, minimum number of drives near the panel and proper orientation of longwall faces reduce the occurrences of bump. The methods likely to destress the immediate roof could further ease mining of bump-prone seam, the most important of which have been: (a) working of protective seam, (b) destressing by drilling hole, (c) volley firings, and (d) water infusion. It is further recommended that destressed seams be extracted fully leaving no residual pillar.

In the light of these observations, the possibility of working thick Dishergarh seam to full height by longwall or bord and pillar even with stowing appeared unsafe and required a special method of mining and system of support.

STUDIES FOR SUITABLE METHOD OF MINING

A number of laboratory studies were undertaken in 1975 and 1980 to find a suitable method for full-seam extraction under a roof thickness of 17 and 31 m. The feasibility of inclined slicing of Dishergarh seam with a thin coal band 0.30 m thick as an artificial roof for lower slices in descending order where seam thickness was 4.8 m was studied first. Extraction of full-seam height to 3 m in a single pass under an immediate roof 31 m thick was analysed next. The third proposition was the composite slicing of Dishergarh seam wherein the full-seam thickness of 4.8 m could be worked in two 2 slices in ascending order, the bottom slice being worked in conjunction with stowing followed by the top section caving over the sand floor.

FEASIBILITY OF WORKING IN SLICES

The experiment [5] was based on borehole information adjacent to 8A panel of Chinakuri Pits No. 1 & 2 (Fig. 4) according to which the immediate 6 m roof has intercalations followed by a coal streak and 9 m thick sandstone. The formation above 20 m was thick massive sandstone likely to work as the main roof. The immediate 17 m roof formation was apparently massive but weak under tension and was likely to cave in after 44 m advance of the face. The

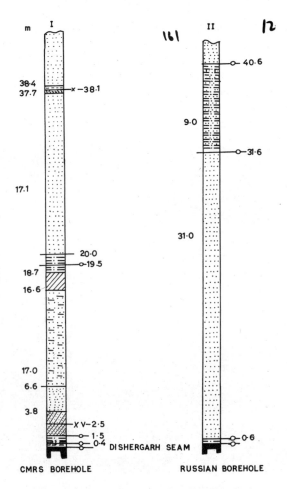

Fig. 4. Strata section near 8A Panel of Chinakuri.

face could be supported in this case by 80 tonne goaf edge chock and 40 tonne props offering overall support density of 60 tonnes/m². Recurring falls were observed after 10 to 15 m interval.

Extraction of the bottom slice leaving a 30 cm thick coal band was not feasible in light of the flowing nature of coal which has a tendency to cave in development galleries. The thin coal band was found to cave in even over a 60 cm span in between the face and tip of the support although bottom slice working after settlement of goaf of the top slice was safe and convenient. It was therefore opined that extraction of full-seam working height in descending order with a coal band as an artificial roof was not suitable for this condition.

FEASIBILITY OF WORKING TO FULL HEIGHT IN SINGLE PASS

This trial [6] was based on borehole information collected by Soviet Experts in the western sector of the mine where thickness of the immediate roof measured 31.0 m with only 0.6 m shale as false roof (Fig. 5). The first major fall with a 31 m thick immediate roof was observed after 80 m face advance in model studies. Subsequent falls occurred after an 18 to 20 m interval. The face during

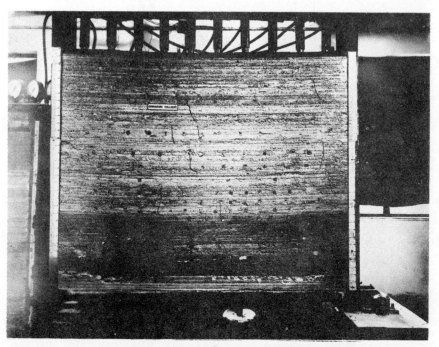

Fig. 5. Model depicting major fall with 31 m thick immediate roof.

caving was under dynamic loading and even 125 tonnes/m^2 support resistance caused 6% roof convergence within the working. The total roof up to 31 m thickness caved in causing a shock wave like a bump. Fragmentation of the immediate roof mass was very poor.

Studies by the Soviet team confirmed thickness of the immediate roof as 17 to 31 m and the first major caving after 54 to 88 m span [7]. Secondary caving was envisaged at a 10 to 18 m interval. The expected maximum load on the support for 31 m immediate roof thickness was 270 tonnes/m^2 with 4 m wide face area for the first major fall and 105 tonnes/m^2 in the subsequent caving. A support of this capacity at that juncture was available neither in the Soviet Union nor elsewhere in the world and hence studies were further extended to proposition 3.

FEASIBILITY OF COMPOSITE SLICING OF SEAM IN ASCENDING ORDER

The seam was proposed to be extracted in slices; firstly, working the bottom 2 m section with stowing followed by the top 2 m section with caving with longwall retreating method. The proposition was based on observations world over and also in Chinakuri mine where longwall was a comparatively safe mining

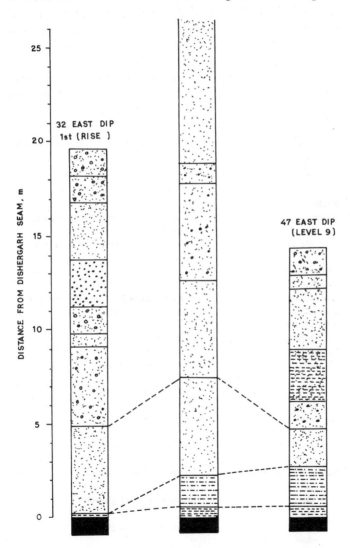

Fig. 6. Strata formation near experimental panel.

method in conjunction with stowing. Extraction of top slice over the stowed goaf was by and large free from bumps because the destressing effect under the influence of bottom section working acted as a protective seam. Laboratory studies revealed increased convergence and reduced span of immediate roof during upper slice workings, when the bottom slice was worked with stowing. This study was based on detailed geological observation of cleats, joints and borehole section recovered from three boreholes drilled in the eastern sector experimental panel around 32 dip level 1, 47 dip level 11-1/2 and 47 dip level 9 junction. The formation (Fig. 6) along these boreholes [8] appeared strong but was full of weaknesses and discontinuities as summarised in Tables 3a and b. The formation on the whole appeared strong with compressive strength varying from 490 to 725 kg/cm^2 and RQD of 97%, but given the presence of a shale streak and mica intrusions the immediate roof thickness appeared favourable for caving.

Table 3a. Qualitative details of boreholes drilled at 47 East Dip 11-1/2 level at Chinakuri Pits 1 & 2

Run horizon, m	Formation	Remarks
0.0 – 0.38	Shale	—
0.38–0.55	Shale	Carbon/graphite streak along breaking surfaces
0.55–0.60	Intercalation	Carbon/graphite streak along breaking surface
0.60–0.61	Shale flakes	—
0.61–0.84	Intercalation	—
0.84–1.35	Intercalation	Carbon streak along breaking surface
1.35–1.39	Intercalation	Carbon/graphite streak
1.39–2.32	Intercalation	—
2.32–2.47	Sandstone (fg)	Shale streak
2.47–4.10	Sandstone (fg)	—
4.10–4.15	–do–	Carbon/graphite streak
4.15–4.68	–do–	—
4.68–5.11	–do–	Carbon/graphite streak
5.11–5.44	–do–	—
5.44–5.56	–do–	Shale streak
5.56–7.23	–do–	—
7.23–7.43	–do–	Intrusion of quartz pebble
7.43–7.60	–do–	—
7.60–7.71	Sandstone (fg) white	Intrusion of mica particles
7.71–8.21	–do–	—
8.21–8.27	–do–	Intrusion of mica particles
8.27–8.59	–do–	—
8.59–8.63	–do–	Intrusion of mica particles
8.63–9.44	–do–	—
9.44–9.57	–do–	Intrusion of mica particles
9.57–9.58	Shale flakes	—

(*Contd.*)

Table 3a (*Contd.*)

Run horizon, m	Formation	Remarks
9.58–10.50	Sandstone (mg) white	—
10.50–10.62	–do–	Intrusion of carbon streak/shale streak
10.62–10.83	Sandstone (fg) white	Intrusion of carbon/graphite and shale streak
10.83–11.36	–do–	—
11.36–12.66	–do–	—
12.66–13.45	Sandstone (mg) grey	—
13.45–14.07	–do–	Shale streak occasionally
14.07–14.70	–do–	Carbon/graphite intrusion
14.70–14.89	–do–	—
14.89–14.97	Sandstone (mg) grey	Carbon/graphite intrusion
14.97–15.15	–do–	—
15.15–15.45	–do–	Shale streak
15.45–16.03	–do–	Carbon/graphite intrusion
16.03–16.75	–do–	—
16.75–17.05	–do–	Intrusion of mica particles

Table 3b. Qualitative datails of the core samples drilled at 44 East Dip, 9th Level Junction at Chinakuri Pits 1 & 2

Run horizon, m	Formation	Remarks
0.0–0.18	Shale	Traces of carbon/graphite intrusion
0.18–0.57	–do–	—
0.57–1.66	Intercalation	Shale prominent
1.66–1.83	–do–	Traces of carbon/graphite intrusion at breaking surface
1.83–1.86	Shale flames	—
1.85–2.49	Intercalation	Intrusion of mica particles and carbon
2.49–3.01	Sandstone (mg)	—
3.01–3.03	–do–	Traces of carbon/graphite
3.03–3.20	–do–	—
3.20–3.21	Shale flames	—
3.21–3.29	Sandstone (mg)	—
3.29–4.00	–do–	Traces of carbon along breaking surfaces
4.00–4.76	–do–	Heavy concentration of carbon/graphite along breaking surfaces
4.76–5.16	–do–	—
5.16–5.80	Sandstone (fg) brownish	Traces of carbon/graphite and shale streak
5.80–6.15	–do–	Traces of carbon intrusion
6.15–6.24	–do–	—
6.24–6.90	–do–	Fine shale streak
6.90–7.20	–do–	Traces of carbon intrusion and fine shale strata

Table 3b *(Contd.)*

Run horizon, m	Formation	Remarks
7.20–7.46	–do–	Heavy concentration of carbon/graphite along breaking surface
7.46–7.60	–do–	—
7.60–8.59	Sandstone (mg)	Traces of carbon/graphite intrusion
8.59–9.52	–do–	—
9.52–10.50	–do–	—
10.50–10.58	–do–	—
10.58–10.93	Sandstone (fg) brownish	Traces of carbon/graphite intrusion
10.93–11.17	–do–	—
11.17–12.60	–do–	Traces of carbon/graphite intrusion
12.60–12.85	–do–	Heavy concentration of carbon/graphite intrusion and mica particles
12.85–13.12	–do–	Traces of carbon/graphite and mica particles
13.12–14.72	Sandstone (mg) grey	—

First prominent horizon, confirmed by both boreholes, was 7.60 m and the formation in the first borehole highly favourable for caving in view of RQD, inclusions and disturbances. Tensile strength along the bedding plane was low—10 to 18 kg/cm²—indicating a well-laminated formation with weak contact along which a bed separation was expected with 4-5 times the working height while working the bottom slice with stowing. Tensile strength across the bedding plane was consistently high for all beds indicating a larger overhang. The first major fall extending up to 5 times the working height was envisaged after 40 to 50 m face advance. The fall was likely to be en masse leading to a shock wave or a phenomenon like a bump. This could be eased by induced caving of the roof and installation of effective support resistance. Recurring falls were expected after a 10 to 15 m interval associated with 4 to 5 m overhang. This might need induced caving in the early stage to reduce chances of shock loading.

Necessary support resistance for the situation was of the order of 225 tonnes per running metre for a working height below 2 m.

BEARING STRENGTH OF SAND FLOOR

Sand-pack during bottom slicing was to act as the floor for top slice caving. In the first trial it was suggested that a thin band of coal be left over the sand-pack but for all practical purposes this was treated as being crushed and pulverised like loose debris. The maximum and minimum supported span of the face was to be maintained at 3.3 and 3.9 m respectively with 60 cm web thickness. The support resistance for this proposition was 55 to 60 tonnes per m². Sand as a floor had only 1.25 tonnes bearing capacity for a 15 cm base diameter which could be increased to 10.5 tonnes by increasing the base to 45 cm diameter. The same base 15 cm deep within wet sand had a bearing capacity of only 20 tonnes. A rectangular sleeper of 1.5 m × 0.25 m × 0.08 m could sustain 45 tonnes load before penetration exceeded 15 cm. It thus appeared that the biggest bottleneck in caving over the top slice on the sand floor was the requirement of broad-based support with floor bearing pressure less than 10 kg/cm². Extended base shields with well-distributed floor stress below the above limit could serve the purpose with modified provision of induced caving and anti-shock valve.

SLICING LAYOUT OF DISHERGARH SEAM

The composite slicing of Dishergarh seam involving extraction of bottom section with stowing and top section caving over the sand floor was recommended to take advantage of protective seam working effect on bumps and degassing of bottom section coal containing a high percentage of gas. The amount of sand required for the full seam working height was reduced to half with stowing while keeping the support resistance within the limit of available range.

Mining of the bottom section of Dishergarh seam with stowing and individual props—friction and timber—has been successful in 11 panels using shearer as the coaling machine. It is suggested that the bottom section be extracted to 1.8 to 2 m height by the method adopted earlier at Chinakuri Pits No. 1 & 2. The layout of the bottom lift face (Fig. 7) was between level 11-1/2 and level 9 with a twin entry layout. Stowing of the top and bottom gateroads was proposed to ensure full stowed goaf for top section caving. Transport of coal from the face was by way of level 12 through stage loaders.

The top slice face is to be started after 100 to 150 m advance of the bottom slice face driving a fresh set of gateroads over levels 9 and 11-1/2 stowed floor up to 40 m in advance (Fig. 8). The face is to be supported by a 4-legged broad-based shield offering floor resistance below 10 kg/cm² and support resistance 50 to 60 tonnes/m². Coal is to be extracted by DERD up to 2 m height leaving 0.5 to 0.8 m coal band as a floor.

This proposition is likely to have the following merits:

1) Working height of different lifts reduced to within 2 m.
2) Mixing of sand to coal obviated altogether.
3) Advance of conveyor and support made convenient.
4) Support more effective due to the presence of coal mass.

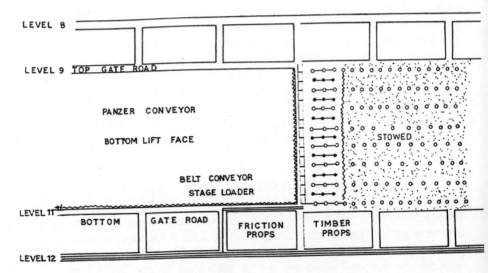

Fig. 7. Bottom section face layout.

BUMP PREVENTION MEASURES

The bottom section working in conjunction with stowing in 150 m long face was likely to cause convergence of immediate roof by 10%. Even a narrow working spread across three pillars only caused 5% convergence within 15 m of face area. Convergence was likely to destress the roof as well as loosen top section coal. This required close stowing and leaving of certain percentage of timber props within the goaf. This would ensure stability of top section coal

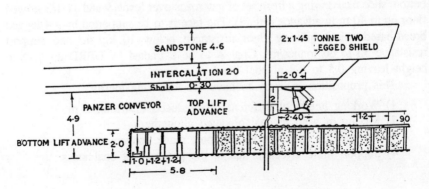

Fig. 8. Top lift face with respect to bottom lift face.

to be worked in the second lift while the advantage of protective seam working for roof caving is retained for future safety.

The top slice face is to operate over the sand-pack, destressed roof and degassified coal seam. The proposition has all the parameters favourable for convenient caving of the top slice face. In light of observations conducted by CMRS as well as the PSWETMETPROMEXPORT, it was further suggested that induced caving of the roof (Fig. 9) be undertaken so as to reduce overhang, and avoid shock wave and air blast. The proposition is likely to produce over 1000 tonnes of coal per day including 300 tonnes from stowing face and 700 tonnes from caved top slice face with favourable economics.

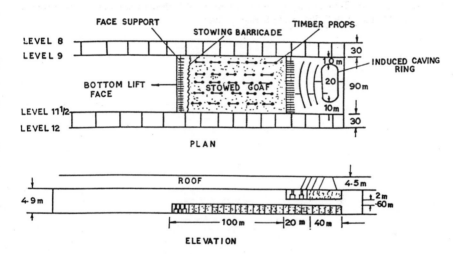

Fig. 9. Induced caving of immediate roof.

ACKNOWLEDGEMENTS

Investigations were conducted at the Central Mining Research Station, Dhanbad with financial assistance from Eastern Coalfields Ltd. The second experiment was conducted in close collaboration with Mr. Vector Chekhov Gerokhov of VNIMI, USSR. The third programme was initiated by Mr. S. N. Singh, Chairman-cum-Managing Director, Eastern Coalfields Ltd., and the assistance of Chinakuri mine management for this study, particularly of Shri R. Srivastava, is thankfully acknowledged.

Model and field experiments were conducted by CMRS team. Among them, the contributions of M/s. M.A. Rafiqui and B.K. Dubey are gratefully acknowledged.

REFERENCES

1. Singh, T.N., M.A. Rafiqui and B. Singh. Bumps in Coal Mines, A paper presented to MGMI, 1969.
2. Saluzstowice, A. 1963. The problem of rockburst in mines. *Int. Mining Congress, Salzburg*, 1963.
3. Ivanov, B.M. 1961. Approximate methods of determining hazard of coal burst. *Ugol*, vol. 36, Sep. 1961.
4. Jacques *et al.* 1964. Instability of mine workings, bumps, rockbursts in the floor and generalized collapse. *4th Int. Congress on Strata Control & Rock Mechanics,* New York, May 1964.
5. Singh, T.N. and B. Singh. Investigations into inclined slicing of bump prone Dishergarh Seam of Chinakuri 1 & 2 Pits. CMRS Investigation Report, 1975.
6. Singh, T.N. and B. Singh. Investigation into caving of bump-prone Dishergarh seam to full height. CMRS Investigation Report, 1979.
7. Vector Chekhov Gerokhov. Personal Communication, 1979.
8. Singh, T.N. *et al.* Technical report on top lift caving of Dishergarh seam over stowed bottom lift at Chinakuri 1 & 2 Pits. CMRS Report, 1985.

13

ROCKBURST PREVENTION SYSTEM INTRODUCED IN THE OSTRAVA-KARVINA COALFIELD AND ITS EFFICIENCY

František Bláha and Adolf Škrabiš

Scientific Coal Research Institute of Ostrava-Radvanice, Czechoslovakia

INTRODUCTION

Rockbursts represent one of the most serious natural phenomena which adversely affect particularly the safety and technology of mining as well as the economy of coal and ore production. Their number and seriousness conspicuously increased in the early 70s in the Ostrava-Karviná Coalfield because of continual worsening of mining and geological conditions, increase in depth of mining and adverse changes in rock mass geomechanical properties.

Rockbursts constitute nowadays a very serious problem in the Ostrava-Karviná Coalfield. According to the forecast, we may expect that some 50% of the total tonnage will be mined in coal seams prone to rockbursts by the end 80s.

BASIS FOR SOLUTION OF ROCKBURST PREVENTION IN CZECHOSLOVAKIA

The conception of rockburst prevention in Ostrava-Karviná Coalfield is briefly explained by the definition of rockburst which, according to present knowledge, is as follows:

Rockburst represents a fracture in the rock mass which results from mining activity during which elastic deformation energy is suddenly released and transformed into work with serious impact on mine workings, mine machinery, and lives of workmen.

A precisely defined set of activities and measures, the so-called "fight against rockbursts", termed in this paper the "rockburst prevention system" has been developed in Czechoslovakia for the recognition of rockburst risk, prevention of its occurrence, and conceptionally includes what is explained in the scheme below:

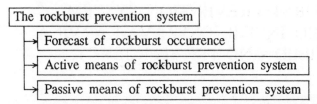

The forecast of rockburst occurrence includes a set of activities which allows one to determine the proneness of the rock mass to rockbursts so that rockburst risk can be classified in planned and active operating mine workings and adequate means of rockburst prevention system selected for application.

Active means of rockburst prevention system represent a set of activities, measures, and equipment applied to counteract rockburst risk.

Passive means of rockburst prevention system represent a set of activities, measures, and equipment applied to mitigate the consequences of rockbursts.

CONCEPTION AND PRACTICAL APPLICATION OF FORECAST OF ROCKBURST OCCURRENCE

The forecast of rockburst occurrence is divided (also according to the presently valid regulations) according to the following scheme.

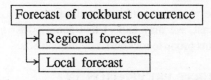

Regional forecast of rockburst occurrence is an activity which allows one to determine the proneness of the rock mass—or parts thereof—to rockbursts. This activity results in the determination of whether a rockburst can originate in the particular deposit or its parts or not.

The forecast is carried out during the phase of deposit exploration and is enhanced with more precision during individual phases of exploitation, i.e., planning, design, and mining. Coal seams, where a rockburst of arbitrary strength has already occurred are classified as dangerous because of rockbursts. Also seams which simultaneously fulfil the following conditions are classified in this category:

— Depth below the surface 300 m or more.

— Elasticity exceeds or equals 70%.

— Complex index of proneness to rockbursts, which comprises evaluation of the influences of depth, structure, and properties of the rock mass, exceeds or equals 3.

Local forecast of rockburst occurrence is an activity which allows one to classify the mine workings according to the degree of imminent rockburst risk and to identify places, where a rockburst can occur. It is carried out during the planning, design, and operational phases of mine workings. According to the local forecast the mine workings operated in seams, which are dangerous because of rockbursts, can be classified in the first, second, or third class of risk. The local forecast relies on the assessment of operation of the particular working in space and time.

ACTIVE MEASURES OF ROCKBURST PREVENTION SYSTEM

Active measures can be categorised (also according to presently valid regulations) according to the following scheme:

Active means of rockburst prevention system
→ Strategic means of rockburst prevention system
→ Tactical means of rockburst prevention system
→ Operative means of rockburst prevention system

Strategic means of rockburst prevention system are aimed at preventing rockburst occurrence and restrict the need for applying other preventive means because of reduction of stress in the rock mass and thus allow one to operate a mine working in a seam with imminent rockburst risk by means of current mining procedures.

Tactical means are defined in the projects of measures to prevent rockbursts and in the technological process and rely on the classification of mine workings according to natural and technical conditions of operation of mine workings and possibilities of their realisation.

Operative measures supplement individual projects of preventive measures against rockbursts and technological processes and rely on the results of local forecasts, assessments of natural and technical conditions of operation of mine workings and possibilities of their realisation.

PASSIVE MEANS OF ROCKBURST PREVENTION SYSTEM

These means principally rely on an artificial release of the rockburst in the absence of workmen and in the protection of workmen in cases when unexpected rockbursts occur.

The most frequently used passive means of rockburst prevention system in Czech coal mines are the following:

— special shotfiring technologies;
— reduction in number of workmen;
— appropriate mine supports;
— prevention of access in mine workings, etc.

ROCKBURST PREVENTION SYSTEM IN LEGISLATION, ORGANISATION AND MANAGEMENT IN OSTRAVA-KARVINÁ COALFIELD

The principal legislative standard for the definition of the rockburst prevention system in Czech mines is the Safety Regulation issued by the Czech Mining Inspectorate, Prague, in 1971. The Regulation of the District Mining Inspectorate, Ostrava, issued in 1981, is linked with the former, and the Instructions issued by the Scientific Coal Research Institute, Ostrava-Radvanice linked with the latter. All the local regulations are due for review in 1988.

In the practical realisation of the above regulations, the following bodies are active:

— General commission of rockburst prevention set up at the General Direction of Ostrava-Karviná Coal Mines. The commission's task is to assess principal concepts of organisation, management, and realisation of the rockburst prevention system in individual mines.
— Disaster prevention commission set up at the General Direction of the Ostrava-Karviná Coal Mines. Its tasks include assessment of particularly complicated situations and investigation of individual disasters. Conclusions of this commission are recommendations to the Chief Engineer of respective collieries.
— Special organisation, which is independent, within the concern Ostrava-Karviná Coal Mines. Its principal tasks are to apply the latest knowledge obtained in scientific research in routine mining practise, to lead methodically the personnel of individual collieries, to approve the methods of mine working operations in the most complicated conditions, and to control the correctness of realised measures of rockburst prevention system.
— Geomechanics department at individual collieries. Its main task is the realisation of rockburst prevention system in the respective mines. The department heads are responsible directly to the Chief Engineer of the colliery.

Scientific knowledge on which the preparation of legislative regulations and the activity of rockbursts relies is systematically developed and supplemented in individual relies is systematically developed and supplemented in individual research and development projects, which are first tested at the Scientific Coal Research Institute, Ostrava-Radvanice.

CONCLUSIONS

An efficient system of rockburst prevention has been developed in Czechoslovakia at present. However, the solution of these serious mining problems cannot be considered complete and definite. In spite of all applied measures, dangerous rockbursts continue to occur in some cases. In addition, the application of measures of rockburst prevention system means in every case an increase in costs per tonne of mined mineral. Therefore, the full attention of research will be dedicated to these problems in future also.

14

FORTY-YEAR EXPERIENCE IN COMBATING ROCKBURST HAZARDS IN DISHERGARH SEAM

R.D. Singh[1] and D.P. Singh[2]

[1]Kothagudem School of Mines
[2]Banaras Hindu University

INTRODUCTION

In India, mining in Dishergarh coal seam is known to be beseı by problems of bumps, besides the problems of high emission of gas and high virgin rock temperature. Some of the worst disasters, explosions and rockbursts have occurred in the Dishergarh seam.

Mining in Dishergarh seam commenced some 50 years ago and since then a variety of techniques have been tried, particularly with a view to mitigating the risk of bumps, and some newer techniques are contemplated as an improvement over past practises. This paper presents an appraisal of experiences gained thus far and the direction in which developments are likely to take place to establish a safe and economic method for exploitation of this seam in years to come.

GEOLOGICAL AND GEOMECHANICAL APPRAISAL

The Dishergarh seam occurs in the Raniganj Coal Measures of the Damuda series of the Gondwana system of Upper Permian age. The associated strata are mainly sandstones (to 88%). A borehole section through Dishergarh seam is shown in Fig. 1.

The sandstone associated with Dishergarh seam has a compressive strength to 1,000 kg/cm^2 at right angles to bedding planes and the following composition: quartz 45-82%, feldspars 28-43% and matrix/cement 11-16% (Singh, 1970).

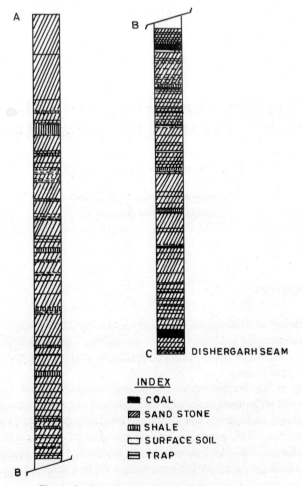

Fig. 1. Section of strata through Dishergarh seam.

The Dishergarh coal seam (3-4.8 m thick) forms an important reserve of good blendable coal and covers a large area along the Grand Trunk road in the vicinity of Asansol (Fig. 2), with an estimated total reserve of about 520,000,000 tonnes (proved, indicated and inferred).

The Dishergarh seam is highly prone to bumps; its dry energy index (W_{ET}) is about 10 (Singh, 1981). It has a compressive strength of 223.01 kg/cm² at right angles to bedding planes and 146.51 kg/cm² parallel to bedding planes.

Figure 3 shows the strength profile of strata associated with this seam at Chinakuri Pits No. 1 and 2 Colliery. The floor is also hard with compressive strength as high as 900 kg/cm². The seam is highly gassy, dry and dusty. Figures 4 and 5 show the stress-strain relationship for Dishergarh seam

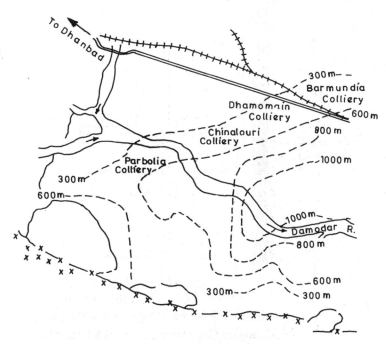

Fig. 2. Map showing floor contours of Dishergarh seam in western sector of Raniganj coalfield to the bedding planes.

coal, roof and floor rocks respectively (Singh, 1965). It can be seen that the stress-strain relationship is linear in all cases. Particularly at a higher range of loading, since stress-causing explosive failure is reached, there is almost no yield of coal. It is obvious, therefore, that the coal (which is the weakest member) will fail suddenly and result in bump, if stresses are permitted to develop above the critical limit.

EARLY EXPERIENCES—A REVIEW

Exploitation of Dishergarh seam in the early years around the 1930s started in all the collieries on bord and pillar system as practised in other coal seams in Indian coalfields. During development work, no problems were encountered except for small coal bursts from the sides of galleries or from the face. However, after pillar extraction began and when the extraction area increased, severe bumps occurred and depillaring with caving had to be discontinued.

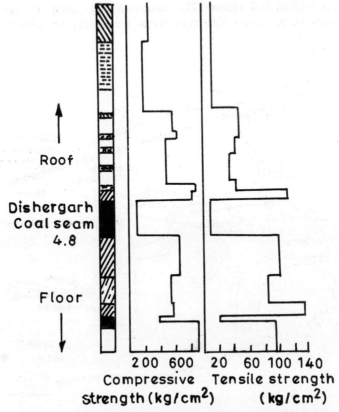

Roof

Dishergarh
Coal seam
4.8

Floor

200 600   20 60 100 140
Compressive   Tensile strength
strength (kg/cm²)   (kg/cm²)

Fig. 3. Strength profile of Dishergarh seam and associated strata.

For example, at Dhemomain Colliery, the Dishergarh seam was accessed by two shafts 350.52 m and 360.27 m deep respectively. The workings in one district extended to a distance of 1,372 m on the dip side of the shaft and attained a depth of 610 m. Extraction of pillars in the extreme dip workings commenced in 1935 and by January 1941 an area 610 m × 152 m had been extracted by open goafing. The goaf was infused and the line of extraction kept level. Up to a goaf width of 152 m, no undue difficulty was experienced in depillaring operations. Thereafter, bumps began to occur. A narrow barrier of solid coal was left to isolate the goaf and depillaring with caving recommenced. The measure proved ineffective, however, and severe bumps began to occur in May 1941; depillaring with caving had to be discontinued (*Mine Disasters in India,* Vol. II, 1981).

Barraclough (1950) has indicated that two kinds of "bumps" have been observed in Dishergarh mines. Shock bumps of great intensity occur in the vicinity of unstowed face worked at depths greater than 300 m, which crush pillars,

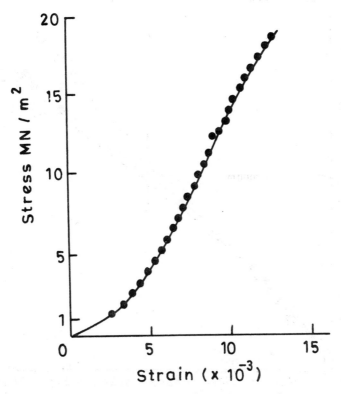

Fig. 4. Stress-strain curve for Dishergarh seam coal.

raise large amounts of coal dust in the air, cause large roof falls and may cause sudden emission of large volumes of inflammable gas from unstowed goaves. Such bumps may or may not be felt on the surface. After such bumps a clear space may be left between the stone roof and coal standing in pillars and spontaneous combustion has occurred in the centre of pillars. The second type of bump occurs in the upper strata and is felt on the surface even to a distance of several kilometres from the focus of occurrence, but fortunately is rarely felt underground. Majority of these bumps have occurred in bord and pillar workings, especially during depillaring operations, which account for 95% of the bumps, of which 73.4% occurred in extraction faces and 26.6% in split galleries.

These observations clearly indicated that the Dishergarh seam could not be worked by caving. Later attempts to extract pillars in Dishergarh seam were always done in conjunction with hydraulic sand stowing.

182

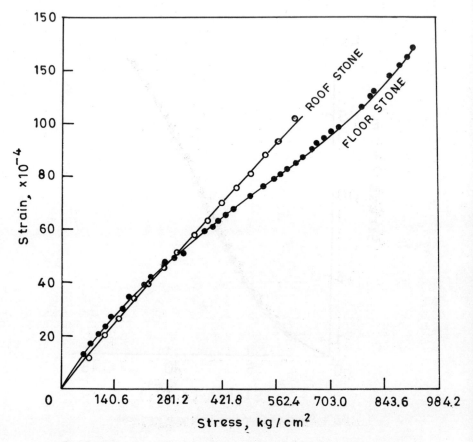

Fig. 5. Stress-strain relationship for roof and floor stone of Dishergarh seam.

EXTRACTION OF PILLARS WITH STOWING

From 1943 onwards, pillars in Dishergarh seam were extracted only in conjunction with stowing but with partial success. The practise followed in the 1950s has already been described by R.D. Singh (1958) and is reproduced below:

Pillars are now extracted in conjunction with solid sand stowing, and as far as possible, a step diagonal line of face is mintained. Each pillar is extracted in slices of 4-5-6 m width, slices being always driven along sand in the destressed zone. Firstly, a 2.1 m of roof coal is taken down and then floor coal is extracted retreating backwards. Close timbering is done by cogs and props, the last row of props being about 1.8-2.1 m from the face. After extracting both the top and the bottom lifts, timbers are withdrawn and the gob is stowed solid with sand. Next, the second slice is taken adjacent to the pack and so on until the whole pillar is extracted.

Figure 6 shows a district in which extraction was coming to a close. On the west side was a barrier about 60 m thick, and on the east side pillars were already extracted up to 8th level. There were only three rows of pillars, each row consisting of three pillars left for extraction on the west side of the haulage line. Previously, severe bumps had been experienced when extracting pillars on the west side whereas work in the east side had been less difficult.

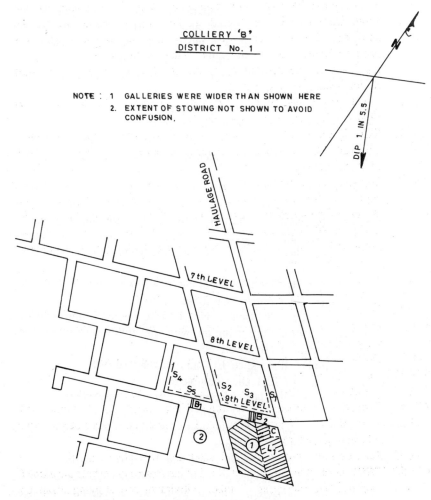

Fig. 6. Colliery 'B' District No. 1.

Galleries were very wide and the roof taken out almost over the whole area except in the main haulage line and about a pillar length in the 7th west level.

A pump was kept on the barrier side. It was decided to shroud the pillar under extraction in sand and then clean the sand to reapproach the pillar and extract it in slices commencing from the destressed zone.

It was decided that a goaf should be formed along the previous goaf line and also that the face should be gradually swung so that the pump could be shifted from the barrier side to the main dip to enable extraction of the barrier pillars above the 6th level. Work was carried out in the following manner:

2.8.52. Double barricade 'B' made in 9th level about 2.4 m apart leaving a passage to pillar No. 1, and galleries on all sides of this pillar stowed solid. Skirted at S_1 and started another skirt S_2. While driving the skirt, only 2.1 m of coal was taken along the roof and sand was packed on the side of the pillar. Made barricade B_1.

Barricade B_0 cleaned for approach to pillar No. 1, and level skirt S_3 for second outlet started.

9.9.52. Level skirt S_3 had become too wide and was timbered. *A thud was reported in coal roof above the 8th level.*

15.9.52. Skirt S_2 joined with level skirt S_3. *When it joined, there was a thud.* Barricade B_0 was cleaned, *widened*, supported, and slice L_1 started. Skirt S_4 started.

18.9.52. Supports put at bottom of skirt S_1. Skirt S_4 was too wide and was supported. Seating begun for pump in 9th level beyond junction.

29.9.52. Water accumulated in Skirt S_5 as a result of stowing beyond No. 2 Dip in 8th and 7th levels. Props in 8th level under weight–two props broke. Slice L_1 in pillar No. 1 going on all right. A 'choukidar' C was left at the corner.

3.10.52. Slice L_1 6 m wide going on along sand. Top 2.1 m of roof coal being taken. *Two cracks developed in roof.*

10.10.52. Slice L_1 joined with sand at No. 10 level. Floor lifting started.

14.10.52. In slice L_1 floor cutting going on, with cogs following behind. Some weight thrown in 8th level.

16.10.52. Opened timber from slice L_1, sand stowing strated.

26.10.52. Sand stowing of first slice completed.

5.11.52. Extraction of pillar No. 1 completed and stowing started.

7.11.52. Extraction of pillar No. 2 started. First started a skirt along sand.

13.11.52. *A prominent roof crack developed in 8th level.* Cogs put on either side and joined by crossbars.

14.11.52. Skirt of pillar No. 2 touched sand and first slice started.

21.11.52. Roof coal taken in first slice of No. 2 pillar. *Prominent cracks developed; roof troubles increased.* Timber withdrawn and stowing arranged.

Figure 7 shows District No. 2 which had an unstowed area of goaf (about 122 m × 488 m) on one side and a dyke about 30.5 m thick on the other; roof along dyke was very bad. This section *had experienced some of the worst bumps,* and the percentage of recovery was rather unsatisfactory. Formerly, pillars used to be split and taken in stooks; but this method did not prove satisfactory and ultimately it was decided to stabilise the area wherever galleries were too wide, and to extract pillars in slices driven along the destressed zone. Operations are described below:

COLLIERY 'B'

DISTRICT No. 2

NOTE : 1. GALLERIES WERE WIDER THAN SHOWN HERE
2. EXTENT OF STOWING NOT SHOWN TO AVOID CONFUSION.

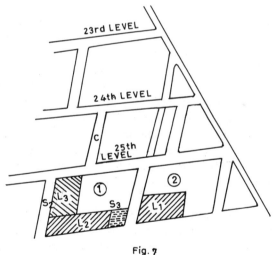

Fig. 7

Fig. 7. Colliery 'B' district No. 2.

15.10.52. Skirt S_1 started, width 2.4 m. Skirt S_2 started. Companion Dip 'C' filled with fallen coal; cleaning started–*severe bumps took place.*
5.11.52. Companion Dip C stowed solid and 24th level half stowed. Skirt S_1 going well. No problems.
10.11.52. Skirt S_1 touched sand and slice 7.6 m width (L_1) started along sand.
14.11.52. Cleaning of 24th level going on. *Bumps around 25th level;* coal cleaning stopped and cleaned area half stowed. *A bump was reported.*

20.11.52. Slice L_1 became very wide (12.2 m). As it had already been driven for a distance of 15 m, decided to take floor coal.

24.11.52. Slice L_1 floor coal taken. Timber withdrawn; ready for stowing. Slice L_2 already driven for a distance of 15 m. *One thud noted.*

27.11.52. Slice L_2: floor coal being taken.

1.12.52. Slice L_2: floor coal taken.
Skirt S_3 going on and *bump reported.*

6.12.52. Timbers withdrawn and stowing of slice L_2 started. *Big bump took place,* most probably due to coal cleaning in crosscut side.

8.12.52. Slice at bottom of skirt S_3 joined with slice L_2. Stowing in 25th level going on.

9.12.52. Stowing of L_2 completed. Slice L_3 started.

10.12.52. *A big bump.* Coal fell in galleries from 20th to 22nd levels.

15.12.52. Stowing in 25th level going on.

21.12.52. 25th level stowed up to companion Dip C, and a skirt started.

1.1.53. Pillar No. 1 finished and stowing started.

5.1.53. All stowed up to level $24^1/2$. Skirt started in east side pillar of 24th level.

The above account shows that bumps continued to occur in spite of stowing the goaf solid with sand emplaced hydraulically and it was realised, at heavy cost of life and loss of coal, that bord and pillar mining was not a suitable method for mining Dishergarh seam. At Dhemomain Colliery a severe bump occurred causing a fall of roof stone 38.1 cm thick over an area of 21.3 m × 6.7 m, killing 12 persons in July 1952. After an enquiry into the accident, the Mines Inspector opined that "the accident lends support to the view that pillar and stall method is not suitable for working of thick seams at depths exceeding 1,000 ft (304 m). Where longwall mining in conjunction with sand stowing has been adopted for extraction of thick seams in virgin areas of deep mines, satisfactory results have been reported". (*Mine Disasters in India,* Vol. II, 1981).

A limited trial to extract pillars on modified longwall method with stowing in descending order gave encouraging results inasmuch as mitigation of bumps was concerned (Singh, 1965). But progress of the face was slow and the output rather low.

Somewhat earlier in England, the South Staffordshire Safe Working of Mines Committee had appointed a sub-committee to review the work done on the occurrence of bumps in South Staffordshire thick coal area and the First Progress Report was published in 1944. The committee summarised its findings as follows:

1) Workings should be laid out so as to prevent coal having to be extracted with old workings on two sides.

2) Wherever possible, the headings should be driven in dead rates.

3) Headings more than 20 yds (18 m) from a parallel heading or old working should be supported by steel arches.

4) The use of coal pillars for supporting the roof in the waste should be avoided where possible, and caving adopted.

5) Laboratory examination of the seam should be made with a view to determining the section in which headings can best be driven.

6) The speed of advance of headings should be adjusted to the conditions; in highly stressed regions one shift in 24 hours.

7) Headings advancing towards each other should be worked from one side only when within 20 yds (18 m) of each other.

8) If possible, a seam above or below the thick area should be worked first.

9) When the thickness of the cave exceeds 400 yds (365.7 m) and the seam is liable to bumps, serious consideration should be given to the selection of a method of working which does not involve in-driving of fast headings in the coal.

Understandably, these findings would have influenced mining systems for the exploitation of bump-prone coal seams in India.

The new method that emerged for extraction of pillars was to take slices from one side only and in a destressed region. Figure 8 illustrates the method of extraction of pillars in Dishergarh seam at Methani Colliery. Extraction was done in conjunction with stowing. The first lift 6 m wide was taken along the roof to take down 2.1 m of coal, the remaining 2.1 m thick floor coal then dug out on retreat and the goaf stowed. But the problems continued. It became clear that bord and pillar method of mining for working bump-prone Dishergarh seam was not the correct choice of mining.

MINING OF DISHERGARH SEAM BY LONGWALL METHOD

The erstwhile Bengal Coal Company Limited introduced longwall mining with stowing in Dishergarh seam at the Parbelia and Sitalpur Collieries in 1944.

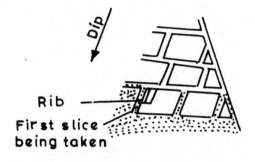

Fig. 8. Part of depillaring district in Dishergarh seam.

Longwalling was limited to the bottom section of 2.1 m only and the top section of 2.1 m was worked by bord and pillar mining. Both single-unit and double-unit faces were worked and mechanical loading was also introduced. Singh (1958), based on Guin (1952), has described the longwall technique as practised those days at Parbelia Colliery:

The seam 4.5 m thick dips at 1 in 7, has a 0.9-1.2 m thick dark shale in the roof, and a hard sandstone floor. Depth of the shaft is 446.8 m and workings are about 610 m deep. The seam is gassy and liable to bumps.

The method of working comprises two phases, namely (i) the bottom 7 ft (2.1 m) of coal is worked by longwall advancing and stowed solid with sand, and (ii) the top 8 ft (2.4 m) of coal is extracted on bord and pillar system and also stowed solid.

After developing a few hundred feet to the dip, a pair of working faces are opened out, leaving about 60 m of barrier from the main haulage road for their protection. The face may be single or double unit. Optimum length of a single unit face is 91 m. For double-unit faces each unit should be 60-76 m.

The faces are arranged along the direction of dip and rise, and advance along the strike. They lie at an angle to the cleats, which gives ease in cutting as the faces become of yielding type. The faces are machine cut and electrically drilled. Shot firing is done with 'permitted' explosives.

On a single-unit face (Fig. 9), there is a shaker conveyor onto which coal is hand loaded, and this transfers coal to a mother-gate conveyor or into tubs standing in the mother-gate. On the double-unit face (Fig. 10) there may be a shaker on the rise face and a belt conveyor on the dip face, both transferring coal onto a mother-gate belt. The dip face is kept in advance of the rise face

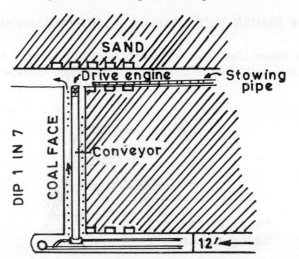

Fig. 9. Layout of a single unit face.

for ease in drainage of stowing water and for protection of electrical gear from water.

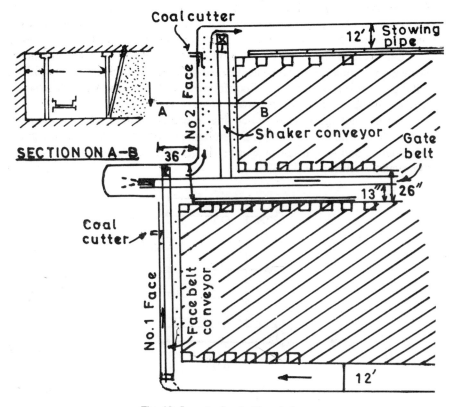

Fig. 10. Layout of a double-unit face

Stowing pipes are laid along the tail and mother-gates. Faces are stowed when a span of 6.4 m has been reached, the adjacent boxings being 3.6 m apart. The face support consists of 15 cm diameter sal props placed 1.8 m apart. Steel props made of 10 cm and 7.6 cm telescopic pipes, with clamps as well as holes and pins, are also in use. Convergence has been found to be 25 mm for every 3.6 m advance of the face. After extraction of the bottom 2.1 m by longwall method, the 2.4 m top coal is developed by bord and pillar system in panels 121 m × 91 m in size with galleries 3 m wide and pillars 27 m square. Pillars are split rise to dip, worked in slices of 6 m width and stowed solid. Actual work consists of scooping some sand from beneath the coal, drilling holes and blasting. To avoid sand mixing with coal, M.S. sheets are inserted in the cuts and coal blasted on top of them. Support consists of cogs

190

placed 3.6 m apart. Sometimes 10-15 cm of roof flakes off suddenly without warning.

The above practise did give considerable relief from bumps but the problem of bumps was not completely solved. Extraction of the top section of 2.4 m coal presented many operational and technical difficulties even though sufficient time was allowed for the sand to settle and consolidate in the bottom slice.

Later efforts have been directed towards the adoption of longwall mining but always in conjunction with hydraulic sand stowing. In virgin areas, longwalling with dip and rise faces along the floor followed by strike faces over the stowed goaf advancing to the rise appeared attractive (Fig. 11).

Double-unit faces with coal-cutting machines, drilling and blasting and hand loading experienced bumps. Severe bumps occurred at the time of goaf consolidation and as a result of blasting in geologically disturbed areas, some bumps also occurred.

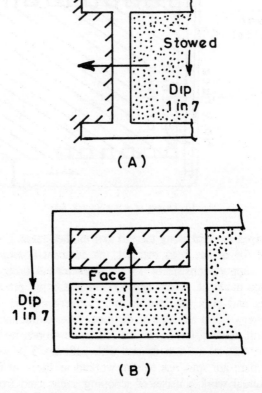

Fig. 11. Working of Dishergarh seam by longwall in two pits.

MECHANISED LONGWALLING

Currently, Dishergarh seam is being mined at pits 1 and 2 in Chinakuri Colliery. At this colliery, the seam has been accessed by two shafts (1 and 2) 612.99 m and 611.12 m deep respectively. The seam is 3.55 m thick and dips at a gradient of 1 in 4.6. The present workings in the mine have reached a depth of 730 m (the deepest coal mine workings in India). Working in these pits in Dishergarh seam, as in other collieries, has also been beset with bumps. Bumps in this colliery have been more frequent in the west side workings, however, in this pit 15 longwall faces have been worked over the period 1963-1984. Table 1 gives the pertinent details of the faces worked. The length of the faces varies from 91 to 182 m. The coal was won by Anderton shearer to a height of 1.5 m from the floor leaving 1.8 m coal in the roof and the goaf was stowed solid with sand emplaced hydraulically. The face was supported by TCR friction props and the minimum and maximum span of the face was 2.4 m and 5.4 m respectively. Experiences over the years are summarised here:

1) Coal bursts of small magnitude occur ahead of the shearing drum when cutting. Coal bursts do not occur when the face is idle, and also no effect of stowing cycle has been discerned.

2) While driving headings, the frequency of coal bursts increases when the heading is within 10 m of an old heading.

3) Over the last five-year period, three high magnitude tremors have occurred which shook the surface but their effect was not felt in underground workings.

4) Road drivage by Dosco dintheader caused falls of 60 cm of roof coal whereas drivage by cutting, drilling and blasting did not result in roof falls.

5) Longwalling with coal cutting, drilling, blasting, hand loading and stowing resulted in the occurrence of bumps but with shearers the occurrence of bumps was almost negligible.

6) In pit No. 3, in the vicinity of the seam worked by bord and pillar method, bumps occurred frequently.

It can be seen from Table 1 that progress of the face varied between 10-24 m per month, i.e., 0.8-0.96 m per day approximately. Such slow progress of the face means that coal was always won in the destressed zone and this would have been the main factor in mitigating the problem of bumps. But such slow progress cannot be considered economic these days. A modern face must produce 2,000 tpd to justify high capital investment and this would mean a daily progress of over 3 m. But "with higher rates of advance of the face, coal will be unfractured and conditions may exist when the static stress is of the order of strength of the coal giving rise to face bumps" (Lama, 1964).

What then was the way out? This question led to scientific research both in the laboratory and in the mine to understand the mechanics of occurrence of bumps in Dishergarh seam, to predict its occurrence and to design suitable

Table 1. Particulars of longwall faces worked in Dishergarh
Seam at Chinakuri Pits 1 & 2 Colliery (1964-1984)

Length of face m	Height worked, m	Length of panel, m	Date of starting extraction	Date of finishing panel	Production per month, tonnes	Average progress per month, m
182	1.70	640	4/63	12/65	8000	20
172	1.80	680	1/64	3/68	9300	21
178.5	1.70	760	12/65	12/68	8500	21
178.5	1.60	490	10/68	—	8000	20
120	1.80	280	12/72	12/73	3200	10
145	1.80	300	6/67	—	8500	24
91	1.80	284	8/64	3/71	5000	15
178	1.80	380	4/69	3/71	8300	16
176	1.85	320	4/64	3/74	4700	12
91	1.80	360	5/74	12/78	4500	17
132	2.20	240	6/77	11/79	4000	10
142	2.10	240	3/73	3/74	6500	20
182	1.98	560	1/78	5/81	6800	18
183	2.20	410	7/79	10/81	4900	11
130	2.10	760	7/81	5/84	6400	14

Source: Data supplied by the management of Chinakuri pits 1 and 2 colliery.

exploitation systems and remedial measures to cope with the situation.

SCIENTIFIC INVESTIGATION AND RESEARCH

Early Work

In India, early workers studied convergence at longwall faces to see the relationship, if any, between convergence and the occurrence of bumps (Guin, 1952; Khanna, 1967). Convergence measurements indicate periodicity of high rates of convergence and the probable occurrence of bumps. Collation and synthesis of field data plotted against time have indicated clear periodicity when bumps could be expected. However, the presence of faults and dykes complicates the situation and bumps could occur quite unexpectedly (Singh, 1958).

Present Work

1) *Laboratory Studies*

The technique developed by Polish workers (Neyman *et al.*, 1972; Szecowka, 1973; Kidybinski, 1981) has been followed in the laboratory to determine the proneness of Dishergarh seam coal to bumps. An index known as the Energy Index W_{ET} is determined by loading a specimen of coal with L/d ratio of 2:1 to 80% of its ultimate strength and then unloading it to zero load. Thus, the total energy imparted to the specimen during loading and the energy released elastically during unloading are determined. The ratio of the elastic energy to the amount of energy remaining in the specimen after unloading is called the Energy Index (W_{ET}). For Dishergarh seam coal the value of the Energy Index has been found to be, on average, around 10 (Fig. 12) which, according to Polish norms, is a highly bump-prone coal. Coal specimens, when saturated with water, gave very low energy indices. The energy index of saturated coal is around 3 (Fig. 13). The inference is that increasing the water content artificially by infusion may reduce the risk of bumps. Also, wetting would be advantageous in reducing the dustiness of the seam and would soften the coal and make it easier to cut. Additionally, the creep characteristics of Dishergarh seam coal were studied (Fig. 14). Even at higher loads, the coal has almost no tendency to creep (Singh, 1981). This suggests that at depths this coal will be highly prone to violent bumps.

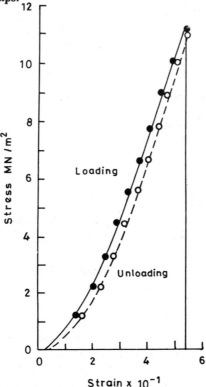

Fig. 12. Stress-strain (loading and unloading) curve for dry coal, Dishergarh seam.

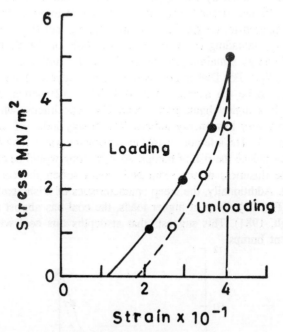

Fig. 13. Stress-strain (loading and unloading) curve for saturated coal.

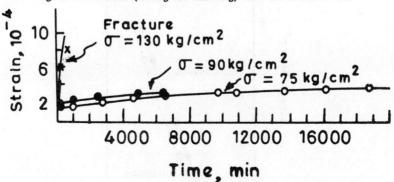

Fig. 14. Creep characteristic of Dishergarh coal at different loads.

2) *Field Studies*

An investigation was made *in situ* to assess the zone of stress concentration at the face. The technique adopted was similar to that used by German and Polish researchers (Brauener, 1971; Neyman *et al.*, 1972). Drill holes were placed at the face and the yield of drilling was measured per metre depth of hole. The position of the drill holes is shown in Fig. 15 and a plot of the yield of drilling against depth of hole is illustrated in Fig. 16. The results indicated that the stress concentration occurred at depths of 7 m inside the coal pillar. This information enabled the management to take stress-relief measures to shift the zone of high stress concentration inside the solid pillar so that extraction could be done in a stress-free zone.

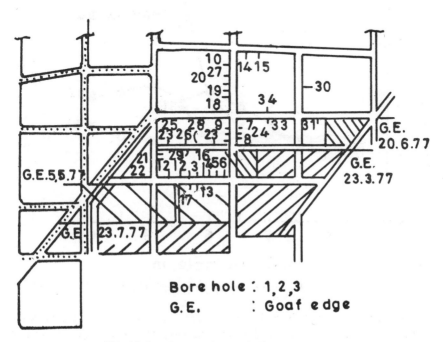

Fig. 15. Location of holes for drilling yield tests in dipillaring district of Dishergarh seam.

To record the onset of bumps and their foci, a Raccal Thermionic Geostore seismic recorder was obtained. It is proposed that data on rockbursts at a number of seismographic stations be collated and analysed with a view to establishing the relationship, if any, between the genesis of rockbursts and the mechanical processes that take place in the rock body during mining extraction. Further, it is intended that seismometers be installed in the vicinity of the mine so that it may be possible to locate the sources of shocks and to determine their energy. The data thus obtained will pinpoint the positions in the mine strata where stresses are approaching the ultimate strength of the rock body and could lead to the occurrence of rockbursts.

Some investigators have studied *in situ* the Dishergarh seam at Chinakuri Pits 1 and 2 Colliery to assess its proneness to bumps. The technique was

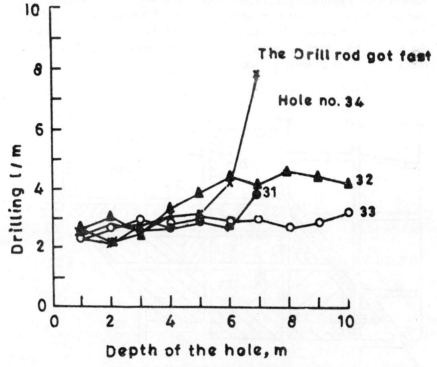

Fig. 16. Yield of drilling versus depth of hole.

to cut a slot in the wall in which jacks were placed and to insert pins in the wall at two horizons above and below the slot. The jacks were pressurised and the closure between them was measured. Subsequently, the pressure was released and the relaxation in the distance between the pins was noted against pressure drop. The results also indicated that the Dishergarh seam was prone to bumps.

EVOLUTION OF POSSIBLE MINING METHOD—"R & D" PROJECTIONS

The accepted method of mining bump-prone coal seams in different countries is the longwall method of mining; thicker seams are extracted in slices preferably with caving in descending or mixed order of slicing. For example, at Wujek Mine in Poland a 7.5 m thick coal seam was worked in three slices in descending order. The top slice along the roof was caved and, leaving a parting 0.5 m thick below the roof, two slices in the lower section were taken in ascending order with stowing. To predict the onset of bumps, seismic recording was done continuously. The roof was induced to cave by blasting long holes. A 100 m longwall face gave a production of 1,200 tpd which would mean a progress of approximately 3.5-4 m per day.

In German mines, test drilling is routinely done to determine whether strata pressure is becoming critical and if so where; the coal could then be won according to the following two principles (Brauener 1972):

1) Face advance is restricted to the rate at which the critical zone usually shifts.

2) if a faster rate of advance is desired, the critical zone's shifting is accelerated by induced stress-relief in the area ahead of the face.

It can be seen that to plan fast-moving faces in order to obtain higher production, techniques for stress-relief in the area ahead of the face should be adopted.

On the other hand, Wilson (1972) based on experiences of gold mining in South African gold mines, has given the following measures for alleviating rockburst hazards:

a) Mining with longwall stopes in order to avoid the formation of pillars.

b) Using rapid yielding props on the stope face to withstand the effects of a rockburst.

c) Using timber concrete supports to control the footwall movements and minimise bed separation in the mining area.

d) Introducing pneumatically placed waste rock in stoped-out areas behind the advancing face. This will reduce the volumetric closure and consequently the rate of energy release. This in turn should reduce rockburst incidence.

e) Designed *"in situ"* stabilising pillars are being left in some mines to control the rate of energy release. Wilson adds that destressing techniques found little use in deep hard rock mines in South Africa.

It is understandable, however, that the depth and conditions in coal mines are far different from those in hard rock deep mines and destressing techniques found useful in Poland, Germany, the USSR, France and Japan may have application in many coal mines. Indian practise has shown that if Dishergarh seam is worked by slow-moving faces with stowing of the goaf, the frequency and intensity of bumps is very much reduced. But if the rate of progress is high, bumps may occur if stowing is not correctly done.

In his reply to Singh's query, Neyman (1972) said that "longwall with hydraulic stowing is safe but only when stowing is correctly carried out". He further added that "such a seam (4.2 m thick) in our conditions, is generally worked in two slices from the top downwards. After the first slice, we put down a timber floor and cave the roof on it. Then we extract the second slice. The distance between the working faces in two slices must be about 40 m to 80 m, depending on the kind of roof and depth of the seam."

In Provence Coalfield, France, relieving hole method is used for the prevention of bumps. This consists of drilling 20-25 m long × 90 mm diameter holes in the solid at 2-3 m intervals from one another (Josien et al., 1982). In Miike mine, Japan, the control of coal bumps is done by driving galleries in the coal seam to regulate discontinuities in the strata and by destress drilling followed by water injection and destress blasting (Osamu et al., 1982).

It has been also suggested that the Dishergarh seam may be worked by longwall caving method and to induce the roof to cave it should be fractured in advance by blasting from the gate roads. If the roof can be fractured in advance, supports with load-carrying capacity of 100 tonnes/m² would be adequate. but the practicability of this technique with fast-moving faces creates problems of its own. Polish researchers advocate blasting of the roof in the goaf to induce caving. In a massive sandstone bed, what would be the length of shotholes to achieve satisfactory results and how much explosive would be needed per blasthole are considerations with their own peculiar legal and safety requirements, the answer to which must be found before adoption of this method to caving faces.

Another approach may be to mine the seam by longwall faces with stowing, but stowing must be done correctly in addition to stress-relief measures taken at the seam level such as borehole destressing and continuous monitoring of the pressure on the face, seismic monitoring etc. Also water infusion of the seam to reduce its strength and effective energy index (W_{ET}) and water injection in the roof to induce caving may merit consideration.

Future R & D activities may be designed along the foregoing considerations.

CONCLUSION

For its quality and reserves, the Dishergarh seam has to be mined and at much greater depths and over larger areas. Severe bumps have occurred in this seam in the past and the potential risk of occurrence of bumps of greater magnitude exists. R & D work supported by practical application of results is the need today to keep mining engineers prepared to face boldly the challenges of the 21st Century and to design the optimum system of mining this seam.

ACKNOWLEDGEMENT

The authors thank the management of Chinakuri Pits 1 & 2 Colliery for making available most of the data on the application of longwall method of mining in Dishergarh coal seam and for the facilities provided for R & D studies.

REFERENCES

Barraclough, L.J. 1950. Hydraulic stowing in India. *Trans. Min. Geol. Metal Inst., India* vol. 45, p. 146.

Brauener, G. 1971. Contribution to the discussion on the paper entitled "Effective Methods for Fighting Rockbursts in Polish Collieries" by Neyman *et al. Proc. of Fifth Int. Strata Control Conf., London.* (London, NCB 1972).

Guin, M.N. 1952. Mining Dishergarh seam at depth, *IMMA Review*, 1, 4, 112-120.

Josien, Jean Pierre *et al.* 1982. Dynamic effects of strata pressure in rockbursts. *Proc. of VII Int. Strata Control Conf., Liege (Belgium), 1982,* pp. 501-521.

Khanna, R.R. Strata movement around a 180 m hydraulically sand-stowed longwall face in the Dishergarh seam. Unpublished lecture delivered at the Winter School on Stratra Central and Rock Mechanics, Central Mining Research Strata, Dhanbad, 8-12 December, 1967.

Kidybinski, A. 1981. Bursting liability indices of coal, *Int. J. of Rock Mech. & Min. Science & Geomech.,* vol. 18, pp. 295-304. (abstract).

Lama, R.D. 1964 Planning of deep mines, *Colliery Engineering,* part I, June 1964, pp. 242-245; Part II, August 1964, pp. 331-333; Part III, September 1964, pp. 242-245.

Neyman, B. 1972. Contribution to the discussion on the paper entitled "Effective Methods for Fighting Rockbursts in Polish Collieries." *Proc. of Fifth Int. Strata Control Conf., London* (London, NCB 1972).

Neyman, B. *et al.* 1972. Effective methods for fighting rockbursts in Polish Collieries. *Ibid.*

Osamu Kimura *et al.* 1982. Study of control of coal burst in Miike Mine. *Proc. of VII Int. Strata Control Conf., Liege* (Belgium), *1982,* pp. 431-448.

Singh, R.D. 1958. A study of deep seam coal mining, *Indian Mining Journal,* Special Issue, pp. 50-59.

Singh, R.D. 1965. A modified method of extracting pillars in a thick seam in Raniganj Coalfield liable to bumps. *Proc. Symp. on Method. of Mining Thick Coal Seams, Min. Geol. Metal, Inst. India. Calcutta.* Paper no. 9.

Singh, R.D. 1970. Techniques of mining thick coal seams in India, pp. 87-109. In: *Mining and Petroleum Technology* edited by M.S. Jones, IMM London. (Proc. 9th Common W. Min. Metall. Conf. 1969, Vol. 1).

Singh, R.D. 1981. Experiences of mining coal seams at depth in India. *Proc. of Asian Mining 81.* IMM London, pp. 159-168

Szecowka, Z., J. Domzal and P. Ozana. 1973. Energy Index of the Inherent Liability of Coal to Rockbursts. *Prace Clownego Inst. Corn Rominikat,* no. 594. Kotowice. (In Polish).

MISCELLANEOUS PAPERS

15

COMPUTER-AIDED ANALYSIS OF IMPACT LOADS ON ROADWAY SUPPORTS DURING ROCKBURSTS

A. Kidybiński

Central Mining Institute, Katowice

INTRODUCTION

In Upper Silesia Coalfield, 65 underground mines produce 192 million tonnes of saleable coal annually from slightly inclined seams of 0.7-8.0 m (average approx. 2.0 m) thickness. More than half these mines experience severe rockburst hazard due to everincreasing depth of working and the appearance of massive sandstone strata in overburden. Breakage of sandstone causes 1400-3700 strong tremors annually, and the seismic energy of a single tremor, affecting the stability of mine openings varies from 1×10^5 J to 1×10^9 J. Traditional roadway support consists of yielding steel arches with open floor section and is extensively damaged due to mine-induced tremors and rockbursts triggered by seismicity. According to statistics, the average length of roadway damaged during a single rockburst depends on the depth of working, as follows:

$$S_d = 23 + 0.1488 \ (D\text{-}400) \tag{1}$$

where S_d is av. length of roadway damaged in metres, and D is depth of mining in metres. Average depth of coal mining in the past decade (at present approx. 560 m) may be expressed by the following equation:

$$D = \exp \ (6.19 + 0.01462 \ (Y - 1978) \tag{2}$$

Where Y is the year. Analysis of seismic energy of mine-induced tremors shows that although the number of tremors is slightly decreasing in time, the average energy of a single shock is rising. Here the line of trend can be expressed by the following equation:

$$E = \exp\ (14.1 + 0.1177\ (Y - 1977)) \tag{3}$$

Where E is energy in J. All these aspects pose a serious problem in designing roadway supports which must resist high dynamic loads. Use of computers for this purpose proved very useful.

ASSESSMENT OF IMPACT LOADS

As reported elsewhere [1], dynamic load on roadway support may be calculated as a kinetic energy of rocks involved in a process of dynamic failure. This energy equals:

$$L = \frac{1}{2}\ Q\ (v_1 + v_2)^2 \tag{4}$$

Where Q is weight of rock moved, v_1 is velocity of rock particles induced by a distant seismic tremor and v_2 is velocity of lumps generated by rapid rock failure. The following formulas are used to calculate these parameters:

$$Q = 0.65\ a\ M\ \left(\frac{2gD}{R} - 1\right) \tag{5}$$

$$v_1 = 2.64 \left(\frac{d}{1.9\ (\ln\ E)^{1/2}}\right)^{-1.55} \tag{6}$$

$$v_2 = \left(\frac{a}{2\ t}\right)\left(\frac{2gD}{R} - 1\right)\left(\frac{g_o}{g_1} - 1\right) \tag{7}$$

where a is the radius of opening cross-section, M is thickness of coal seam exposed, g is overburden density, D is depth of opening from surface, R is compressive strength of a coal seam, d is distance from seismic source to mine opening analysed, E is seismic energy of a tremor, t is duration of dynamic rupture process, g_o is original density of coal and g_1 is density of coal after rupture.

The ten parameters mentioned above may vary significantly for a number of mines, and three among them require special investigations, namely R, E and t. "R" should represent the strength of coal seam as against coal samples, "E" should be a result of assessment of most probable energy of seismic events in a given mining area and may only emerge from long-time seismic recording and statistical analysis, and finally "t" may be evaluated both from laboratory tests and underground seismic records of rockburst cases.

Moreover, there is a random function (d) which resists any forecasting, although some attempts have been made on the basis of local geological studies and statistical analyses of case histories.

Using computer calculations make it possible to alternate parameter values as often as necessary to present the whole range of conditions which may occur in mines, and thus to find general recommendations both for safety and supports manufacturing. Examples of such calculations follow.

COAL RUPTURE ZONE

In order to assess the weight of rock moved (Q) during an anticipated rockburst one should evaluate the size of probable coal rupture zone around the section of mine opening. To this end, mathematical functions of stress distribution in rock surrounding the cavity may be used together with appropriate rock failure criteria. Figure 1 gives an example of such calculations for a depth of mining 560 m which is the average statistical depth of Upper Silesian Collieries.

Similar data are presented in Fig. 2 but for a depth of 700 m. Comparison of these two figures gives a clear idea how the weight of rock engaged in the rockburst process increases with depth.

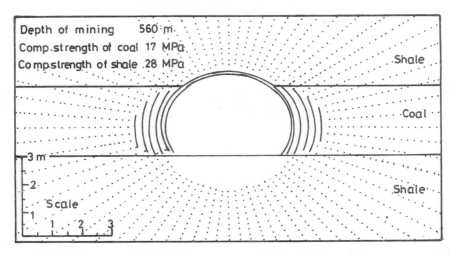

Fig. 1. Coal rupture zone at 560 m depth.

The most severe hazard of rockburst in Silesia is connected with thick seams and roadways with considerable thickness of coal below floor level. In such a case it often happens that 70-80% of the opening volume at the critical section is filled with moved coal masses.

Figure 3 illustrates a coal rupture zone in a thick seam. By taking the initial density of coal as 1.3 t/m³ and final density of coal masses after rupture as 0.9 t/m³, it is easy to calculate part of the opening's volume which may be filled during rockburst.

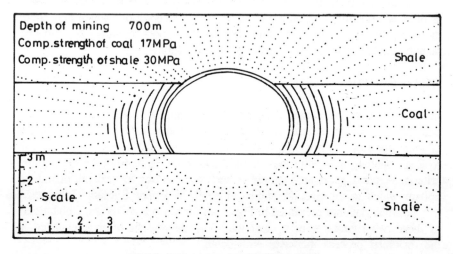

Fig. 2. Coal rupture zone at 700 m depth.

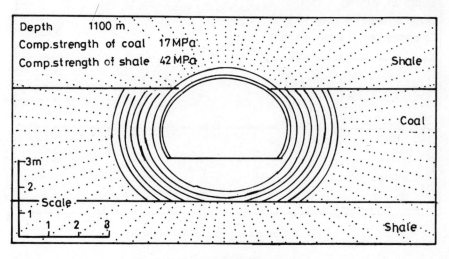

Fig. 3. Coal rupture zone in thick seam.

DISTANCE FROM SEISMIC SOURCE

Energy of impact on supports during rockburst depends on distance of opening

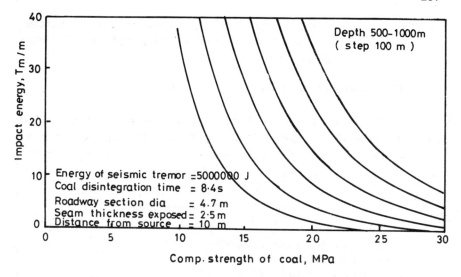

Fig. 4. Impact energy at 10 m distance from seismic source.

considered to the seismic source. Although this parameter is hardly foreseen, it is necessary to know the maximum possible loads for support design purposes.

Curves of impact energy are shown in Fig. 4 for very close distance to the source (10 m). From this graph, it is possible to evaluate for various seams defined by the strength of coal and the required dynamic resistance of supports. When the dynamic resistance of a single steel arch is known, it is easy to calculate the required arch spacing as well as special support reinforcements.

Similar curves are shown in Fig. 5 but for a distance equal to 100 m. It

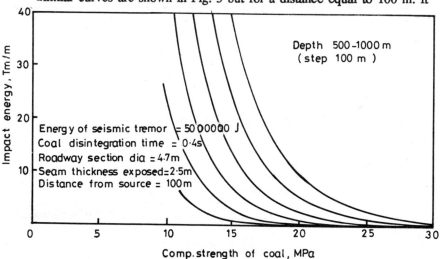

Fig. 5. Impact energy at 100 m distance from seismic source.

208

is important to know from these types of calculations that impact energy decreases significantly only from zero to approx. 100 m distance and for more distant sources remains practically constant.

BURSTING PROPENSITY OF COAL

For stress-generated rockburst the natural bursting propensity of rocks plays a very important role, and that is why several testing methods and classifications

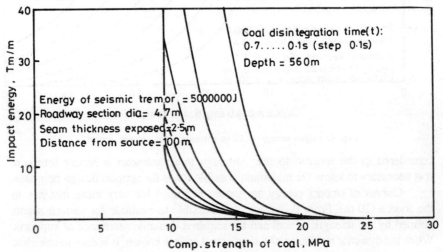

Fig. 6. Effect of bursting propensity (t) of coal on impact energy (at 560 m depth).

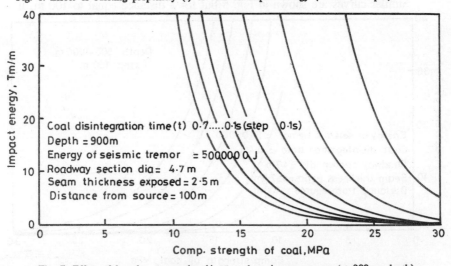

Fig. 7. Effect of bursting propensity (t) of coal on impact energy (at 900 m depth).

have been proposed in the past [2]. For seismicity-generated rockbursts, however (and this type prevails in Upper Silesia), this role is rather limited and bursting propensity is defined as the time duration of dynamic rupture process (t) affecting velocity of mass movement. In Figs. 6 and 7, the effects of t on impact energy are shown for mining depths of 560 and 900 m respectively.

REFERENCES

1. Kidybiński, A. 1986. Design criteria for roadway supports to resist dynamic loads, *Intern. J. Min. and Geol. Engrg.*, no. 4, pp. 91-109.
2. Kidybiński, A. 1982. *Fundamentals of Strata Control Engineering*. Slask Publ. House, Katowice. (In Polish).

16

AMPLITUDE DISTRIBUTION ANALYSIS OF THE ACOUSTIC EMISSION EVENTS RECORDED FROM ROCKS STRESSED TO FAILURE

M.V.M.S. Rao[1], X. Sun and H.R. Hardy Jr.[2]

[1]National Geophysical Research Institute, Hyderabad-500007 (India)
[2]Geomechanics Section, Department of Mineral Engineering,
The Pennsylvania State University, University Park, Pa 16802, USA

INTRODUCTION

Acoustic Emission techniques have been extensively used during the last two decades for carrying out studies on rock fracture in laboratory experiments and also for geotechnical problems, and the state-of-art has been reviewed by Hardy (1972, 1981, 1987). The vast majority of experimental work on acoustic emission study of rock behaviour under varied stress conditions has concentrated till recently on event and ring down counts only. But in the last few years, the development of AE recording and signal processing techniques, and the advent of fast computers to handle large sets of data have been making it possible to not only characterise but also retrieve the complete information contained at AE signals, through multiparameter approach and pattern classification methods. This would be a great advantage for making a detailed analysis of AE events associated with the various stages of microcrack growth (both stable and unstable) that contribute to the gross deformation and failure of brittle rock under compressive loading conditions. We have recorded acoustic emission signals on the magnetic tapes of a videocorder from a few rock specimens that are stressed to failure under uniaxial compression and processed the data off-line using the Dunegan/Endevco Distribution Analyser (Rao et al., 1988). The histogram records of AE event amplitude distribution obtained at different stress intervals during incremental loading till failure have been analysed and an estimate of stress-induced microcrack damage has been attempted on the basis of the distribution of high amplitude AE events in rocks stressed to failure.

EXPERIMENTAL ARRANGEMENT

The MTS facility, used in the current studies, incorporates an electronically programmable, closed-loop electrohydraulic loading system (Hardy *et al.*, 1971) capable of carrying out mechanical compression and tensile tests over a wide range of conditions. The facility also includes equipment for monitoring and recording stress-strain and acoustic emission data during testing. An important feature of the MTS loading facility is the ability to program functions such as load, strain and stroke. In the current studies load control was utilized.

Laboratory experiments were conducted as per the guidelines described in detail by Mrugala & Hardy (1987) using the microcomputer based digital acquisition system (μmac 5000). A functional block diagram of the total experimental arrangement has been shown in Fig. 1 which includes a Sony

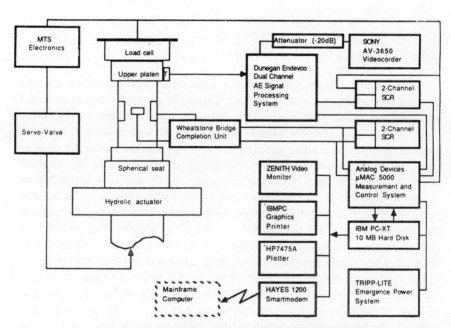

Fig. 1. Block diagram of MTS loading facility and the
PSURML microcomputer-based digital data acquisition system.

Videocorder for recording the acoustic emission signals. The contents of the data tapes were processed off-line by using the standard commercially available Distribution Analyser (Dunegan Endevco Model 920) and its associated equipment (Amplitude Detector-Model 921 A and Signal Conditioner-Model 302 A). The system had the ability to characterize each acoustic emission event that can help in developing a "Signature" for various stages of crack growth or other sub-parts of the emission history. Essentially the Distribution Analyser quantifies each AE burst or event based upon any one of the pre-selected parameters: ringdown counts, amplitude, pulse-width, energy, arrival time difference. In operation, the Distribution Analyser accepts and sorts the AE events into 101 slots (memory elements) with assigned values of 0, 1, 2, ... 99, 100. Then it updates its 101 segment memory with the numerical value of the measured parameter so that, at the conclusion of the test or at an appropriate test increment (as in the present study), the exact distribution of the AE events in the form of a histogram for the selected characteristic can be obtained. the contents of the memory can be directly read on an Oscilloscope and permanent records may also be obtained by connecting an X-Y recorder to the system. A detailed description of the data processing facility and its limitations have been discussed elsewhere (Rao *et al.*, 1988).

RESULTS AND DISCUSSION

Threshold Level Setting of Distribution Analyser

Large number of acoustic emission events (few thousands) have been recorded in each experiment. The amplitude levels of individual events recorded on the videocorder through a 20 dB attenuator were found to range between 0.05 and 1.0 V in most of the rock specimens tested. The total background noise of the Videocorder and the data analysing equipment was found to be of the order of 10 mV making the signal to noise ratio of AE events to be equal to 5. This allowed us to choose 60 dB (0.1 mV) as the lowest threshold level for the amplitude detector of the Distribution Analyser. Trial runs were made initially by setting a gain of 20 dB for the Signal Conditioner to compensate for the 20 dB attenuator that was used during data recording, while playing the data tapes back at selected stress intervals. it was noticed that the memory of the Distribution Analyser was getting saturated fast in the 'NORM' mode to handle the data particularly at stresses close to failure. Attempts have also been made to process the data by selecting the 'LOGARITHMIC' and 'SUM' modes of the Distribution Analyser so that it could also yield the slope of the amplitude distribution ('b' value). Even then it was noticed that the capability of the Analyser was insufficient to process the data acquired at higher stress levels. By gradually increasing the threshold level of the amplitude detector and by reducing the

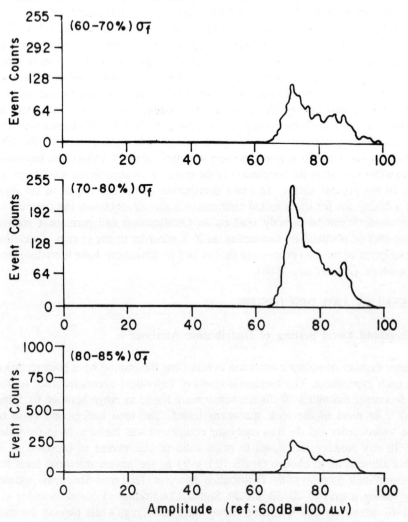

Fig. 2. Differential event amplitude distribution shown in NORM mode
with increased sensitivity during small stress intervals between 60% and 85% σ_f
in schist (C-10B-3) at 75 dB threshold level.

gain of the Signal Conditioner, the capability of the Distribution Analyser was
found to improve considerably. The optimum threshold level was found to be
75 dB for the amplitude detector and 10 dB gain for the Signal Conditioner.
The overall AE activity (number and rate of release of emissions) in all the

DETAILED RECORD OF AMPLITUDE DISTIBUTION

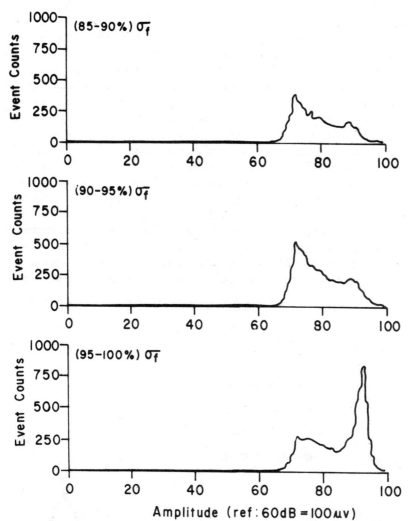

Fig. 3. Differential event amplitude distribution shown in NORM mode with increased sensitivity during small stress intervals between 85% and 100% σ_f in schist (C-10B-3) at 75 dB threshold level.

three end-capped specimens was far below than that of granite and sandstone which did not have end-caps. As a result, attempts were made to process the same data at lower Amplitude Detector threshold levels. The results showed that data processed at 75 dB threshold level gives more meaningful information about the AE activity at stresses that are close to failure.

Amplitude Distribution

In view of the memory limitations of the Distribution Analyser the recorded data had to be processed in steps and not continuously. The data obtained during each stress interval (adjustable) was processed and useful histogram records of amplitude distribution of AE events could be obtained. Since the data was recorded on the magnetic tape, it became possible to get histogram records with increased sensitivity at any desired stress level or during any stress interval. We had particularly done this for the data obtained during smaller intervals of incremental stress at higher stress levels since maximum stress induced microcrack damage (high level AE activity) takes place in the rock just before its failure. Some of the portions of the detailed histogram records of AE amplitude distribution is a Kolar schist are shown in Figs. 2 and 3. These records show that there are essentially two groups of events whose amplitude levels range between 60 and 80 dB (Peak at 70 dB) for group I and 80-100 dB (Peak at 90 dB) for group II. For the purpose of detailed analysis, the events are categorised on the basis of their amplitude data from these records into four ranges, viz. 60-70 dB, 70-80 dB, 80-90 dB and 90-100 dB. The area of each one of these four segments is measured accurately with a Planimeter in every

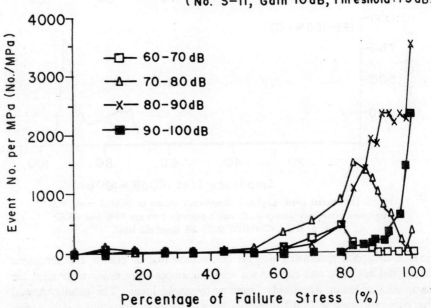

Fig. 4. Stress-induced event count (dN/dσ) versus the percentage of failure stress in sandstone (S-11).

record and the equivalent event count obtained. The event count data thus obtained in these four amplitude ranges per unit rise of stress (dN/dσ) is computed for successive stress interval ranges and the results are plotted as shown in Fig. 4. Events of group I are larger in number at all stress levels up to 90% of σ_f (σ_f is the uniaxial compressive strength), where they become almost equal to those of group II (Fig. 3). When the applied stress increases beyond this limit, group I events are reduced in number (sometimes significantly) while a sharp rise in the number of events of group II is recorded. The total sum of events in both the groups (area of the histogram) at stresses close to failure may apparently reduce and this is attributable to the limitation of the data handling capability of the Distribution Analyser. The volume of data at very high stress levels just before fracture is so large that the equipment is not capable of fully sorting out all the events. This is illustrated more clearly in Fig. 4.

Inferences about Microcrack Damage

Based on the data of AE activity in terms of dN/dσ which is a manifestation of microcrack growth and damage in rock (Costin and Holcomb, 1981), one could perhaps attempt to write down the general equation for the stress-induced crack growth as follows:

$$\frac{dc}{d\sigma} \alpha \left(\frac{dN}{d\sigma} \right)^m \tag{1}$$

where c is the current crack length (or damaged area) and N is the number of events per unit rise of applied stress. It follows that where P is the proportionality

$$\frac{dc}{d\sigma} = P \left(\frac{dN}{d\sigma} \right)^m \tag{2}$$

constant and m is a material property whose value depends on E and σ_f of the test specimen. C should be assumed as an ensemble of cracks instead of a single crack for the present study.

The total change in crack length (or area damaged) at any stress level due to incremental loading is given by

$$\int_{\sigma_i}^{\sigma} dc = \int_{\sigma_i}^{\sigma} P \left(\frac{\partial N}{\partial \sigma} \right)^m d\sigma \tag{3}$$

Where σ_i is the stress level at which microcracks are initiated which is around 50% of σ_f in the present study. It is assumed that failure will occur when the population of microcracks grows to some critical state. This assumption is implicit in many microcrack damage models (Costin & Holcomb 1981).

Failure occurs when

$$\int_{\sigma_i}^{\sigma_f} P\left(\frac{\partial N}{\partial \sigma}\right)^m d\sigma = \Delta K \quad \text{(constant)} \tag{4}$$

It may be mentioned here that contributions to AE activity could also come from other environmental factors such as humidity, temperature, stress corrosion and also the time factor as the specimen is deformed. It is assumed that AE activity arising due to these factors is insignificant in our present study. The above equation implies that if $dN/d\sigma$ is known before the critical state is attained, it is possible to estimate the state of damage or predict the failure stress of the rock.

The AE activity records in Fig. 4 display that the events of group I show decreasing values of $dN/d\sigma$ at stress levels of 80% and $\geq$ 95% σ_f respectively. This could be due to the limitations of the data handling capability of the Distribution analyser or may be attributable to the inherent properties of the rock. So, for the purposes of inferring the microcrack damage, the data of group II events alone i.e. ($dN_3/d\sigma + dN_4/d\sigma$) have been taken into consideration and these are plotted in the logarithmic scale against percentage of failure stress as shown in Fig. 5. The results show that there are 3 distinct regions which show marked changes in the slope of the curve. Region 1 corresponds mainly to the AE activity due to the closure of microcracks/voids in the rock under the application of stresses that are $\leq$ 20% σ_f. The activity in region II almost remains steady (like background noise) and is perhaps attributable to the elastic deformation and the phase that is preparatory to the initiation and growth of microcracks. Region III is found to begin at 80% σ_f and increases very sharply from there till failure. This is noticed to last only till 50% σ_f in the case of sandstone and up to 80% σ_f in schist and dolerite which are end-capped. Significant changes in $dN/d\sigma$ are found to take place only in region III in all the rocks and it is this region which accounts for major damage, as evidenced from sharp increases in other AE parameters also (Rao et al., 1988). These results are completely analogous to the crack velocity-stress intensity factor curves that one obtains during the Creep and Fracture tests (Anderson and Meredith, 1987; Costin, 1987). The stress induced damage which is manifested as AE activity can be qualitatively expressed in terms of $dN/d\sigma$ which is found to attain a peak value only during that stress interval which is very close to failure stress. Taking this as the base value, the data during all the other stress intervals at lower stresses are normalised and the results are plotted against normalised stress as seen in Fig. 6. Taking the damage ($dN/d\sigma$) at failure stress as 1.0, the stress induced microcrack damage at the other stress levels can be estimated

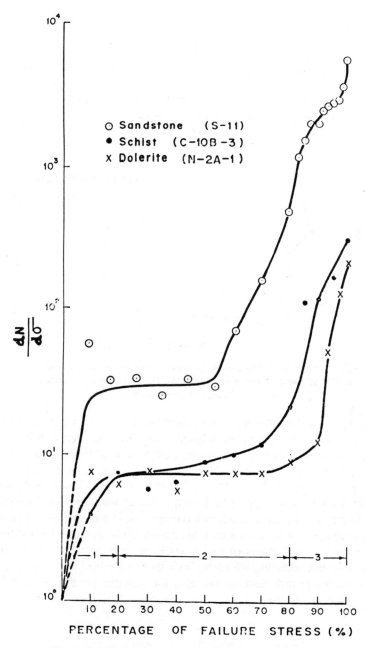

Fig. 5. Stress-induced event count (dN/dσ) of group II events
plotted in logarithmic scale as a function of stress.

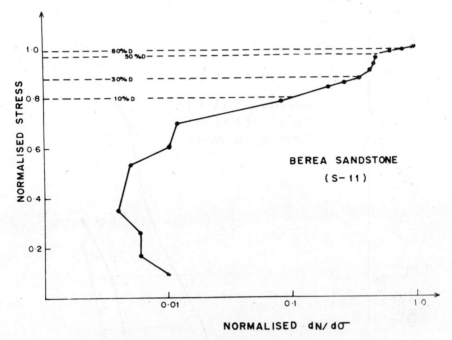

Fig. 6. Normalised values of stress versus stress-induced event count (dN/dσ) of Group II events for sandstone. The stress levels at various stages of damage are also indicated.

in a qualitative manner from this. The results of all the 3 rocks in the present study clearly show that the stress induced microcrack damage at 80% of σ_f is only 10% of the total damage and it can attain a value of 50% only at stresses $\approx$ 95% of σ_f. Also the remaining 50% damage is found to occur only when the stress is raised from 95% to 100% σ_f. These results thus clearly demonstrate that dN/dσ is a very useful parameter for a qualitative assessment of microcrack growth and for estimating the microcrack damage AE monitoring and amplitude distribution analysis of AE events recorded during cyclic fatigue tests and time-dependent crack growth monitoring tests would be more useful since the crack growth is slow and stable in such cases. Such studies can lead to a more precise estimation of microcrack damage and also the possible prediction of failure time and stress of the rock.

CONCLUSIONS

The amplitude distribution of acoustic emission population shows that events recorded at stress levels $\geq$ 95% σ_f of failure stress are reduced in number but have relatively high amplitude and longer duration. Inferences about microcrack damage in terms of dN/dσ of AE events show that the damage is only around 10% at 80% σ$_f$ and attains a value of 50% of total damage at 95% σ$_f$ for all the rocks tested. The analytical equipment used in the present study would perhaps be more ideally suited to make a detailed AE study of slow crack propagation and stress-corrosion cracking in rocks where the emission population is small in number.

ACKNOWLEDGEMENTS

This work was carried out while Dr. Rao was working as a Fulbright Scholar at the Geomechanics Laboratories of the Department of Mineral Engineering, the Pennsylvania State University during 1987-88. The fellowship granted to him by the Council for International exchange of Scholars, Washington, D.C., U.S.A. is gratefully acknowledged.

REFERENCES

Atkinson, B.K. and Meredith. 1987. The theory of subcritical crack growth with applications to minerals and rocks. Chapter 4, in *Fracture Mechanics of Rock*. (ed.) B.K. Atkinson, Academic Press, London, pp. 111-166.

Costin, L.S. and D.J. Holcomb. 1981. Time-dependent failure of rock under cyclic loading. *Tectonophysics*, 79, 279-296.

Costin, L.S. 1987. Time-dependent deformation and failure. Chapter 5 in *Fracture Mechanics of Rock*. (ed.) B.K. Atkinson, Academic Press, London, pp. 167-215.

Hardy, H.R. Jr., 1972. Application of acoustic emission techniques to rock mechanics research. *Acoustic Emission* (eds.) R.G. Liptai, D.O. Harris and C.A. Taro, STP 505, Am. Soc. for Testing & Materials, Philadelphia, Pennsylvania, pp. 41-83.

Hardy, H.R. Jr., 1981. Applications of acoustic emission techniques to rock and rock structures: A state-of-the-art review. *Acoustic Emission in Geotechnical Engineering Practice, STP 750*, V.P. Drnevich and R.E. Gray (eds.), Am. Soc. for Testing & Materials, Philadelphia, Pennsylvania, pp. 4-92.

Hardy, H.R. Jr., 1987. A review of international research relative to the geotechnical field application of acoustic emission/microseismic techniques. *Sixth International Congress on Rock Mechanics*, Montreal, Canada, Aug 30-Sept. 3, 1987.

Hardy, H.R. Jr., R. Stefanko and E. Kimble. 1971. An automated test facility for rock mechanics research. *Int. J. Rock Mech. and Min. Sci.*, 8, pp. 17-28.

Mrugala, M. and H.R. Hardy, Jr. 1987. The development and application of a microcomputer-based digital data acquisition system for use in the Penn State Rock Mechanics Laboratory. Internal Report RML-IR/87-2, Geomechanics Section, Dept. of Mineral Engineering, The Pennsylvania State University, 1987, pp. 1-82.

Rao, M.V.M.S., Sun, X. and H.R. Hardy, Jr. 1988. A Detailed Study of Acoustic Emission Activity Recorded from Rocks Stressed to Failure. Internal Report RML-IR/88-1, Gemechanics Section, Dept. of Mineral Engg., The Pennsylvania State University, 1988, 1-128.